The Value of Science in Space Exploration

THE VALUE OF SCIENCE
IN SPACE EXPLORATION

James S. J. Schwartz

OXFORD
UNIVERSITY PRESS

OXFORD
UNIVERSITY PRESS

Oxford University Press is a department of the University of Oxford. It furthers
the University's objective of excellence in research, scholarship, and education
by publishing worldwide. Oxford is a registered trade mark of Oxford University
Press in the UK and certain other countries.

Published in the United States of America by Oxford University Press
198 Madison Avenue, New York, NY 10016, United States of America.

© Oxford University Press 2020

CIP data is on file at the Library of Congress
ISBN 978-0-19-006906-3

9 8 7 6 5 4 3 2 1

Printed by Sheridan Books, Inc., United States of America

For all who work to improve upon on our understanding of the universe.

. . . there are mysteries that ride on the Sun's rays, majesties in the swirling gases and chunks of matter, and humans will benefit by learning to see other worlds, other events where they are for what they are, as surely as they benefit by having air, water, and soil. The historical struggle, repeated now in ourselves, has always been to get a big enough picture; and now we stand at an exciting place: one world trying to figure out the others.
—Holmes Rolston, III

CONTENTS

PREFACE

I am often asked how I came to work on issues related to space exploration—an unconventional specialization for a philosopher. And, while I have been passionate about space exploration for as long as I can remember, attended Space Camp as a child, and graduated from the same high school as former NASA Deputy Administrator Lori Garver, the actual story is unromantic: One morning, following a night out with friends, I found myself wondering why I had never seen any *philosophy* papers about space exploration. So, on a friend's couch, groggy and dehydrated, I retrieved my laptop and started writing what would eventually become my first publication. Months of research and revisions uncovered a modicum of philosophical work on space. Since then, it has been pleasing to witness and participate in the steady growth and development of philosophy of space exploration. This book provides the first unified, detailed, and substantive exposition and defense of my views on space exploration, especially those pertaining to the absolute and relative importance of *scientific* aspects of spaceflight. My overarching argument is that space science is uniquely epistemically and ethically valuable, and that this value should play a foundational role in thinking about space policy, regulation, and spaceflight objectives.

As anyone who has worked on projects related to space exploration is aware, space research is highly multidisciplinary. This is apparent in obvious ways—the successful planning and implementation of space missions of any kind require individuals from diverse disciplinary backgrounds (engineers, scientists, technicians, astronauts, administrators, politicians, etc.) working in concert to solve varied and difficult problems associated with spacecraft design and construction, mission goals and budgets, mission implementation, and so forth. A similar phenomenon is apparent when it comes to space policy research. Viable and meaningful contributions to space policy discussions are possible only through the acquisition of a broad base of working knowledge related to space exploration.

Engineers and space scientists interested in joining space policy discussions will quickly find they must first learn something about philosophy, political science, or law. Philosophers hoping to join these discussions will quickly find they must first learn something about astrobiology, orbital mechanics, or planetary science. Questions emanating from this book's cynosure—the value of space science—are seldom purely philosophical and must be addressed in ways that acknowledge potential contributions from many other disciplinary perspectives. Even within my home discipline of philosophy, plural expertise is a necessity. Space policy debates draw not only on ethics (including environmental, normative, meta-, and applied ethics), but also from philosophy of science (in both its general and specialized forms), from social and political philosophy, from philosophy of law, from metaphysics and epistemology, and from philosophy of religion. Even philosophy of mathematics offers auspicious portents for space policy! My hope, then, is that this book provides an insightful and instructive model of interdisciplinary research.

I owe particular debts of gratitude to Charles Cockell and Tony Milligan. Charles was the first to welcome me into the (small but growing!) community of scholars who actively research philosophical and other ethical issues raised by space exploration. I met Tony in 2014 at the second of Charles's three *Extraterrestrial Liberty* meetings (held at the British Interplanetary Society in London). Tony invited me to work with him on what would become the edited volume *The Ethics of Space Exploration*, and was kind enough to grant to me the role of lead editor. Readers of Tony's work on space, including his book *Nobody Owns the Moon* (Milligan 2015a), will note a certain synergy between our views on space exploration.

I give many thanks to Wendy Whitman Cobb for her assistance in the construction and interpretation of the regression models I use in Chapter 1 to discuss the impact of funding on degree conferral rates. For commenting on previous drafts of this manuscript, and for discussing questions and issues raised therein, I thank: Susan Castro, Susan Sterrett, Pat Bondy, Charles Cockell, Tony Milligan, Frans von der Dunk, Kelly Smith, Robert Jedicke, and an anonymous referee at Oxford University Press.

For supporting my early work on space while earning my PhD at Wayne State University, I thank: Susan Vineberg, Gonzalo Munévar, Marsha Richmond, Katherine Kim, Travis Figg, Tim Kirschenheiter, Marcus Cooper, John Corvino, Paul Graves, and Barry Johnson.

Since joining the Philosophy Department at Wichita State University I am particularly thankful for the input from: Susan Castro, Susan Sterrett, Brian Hepburn, Noell Birondo, Jeffrey Hershfield, Bob Feleppa, Jay Price, Helen Hundley, Pat Bondy, Nick Solomey, Mark Schneegurt, Ryan Amick, Atri Dutta, and Neal Allen. Many topics treated in this book were discussed in my Summer 2018 honors seminar on the philosophy of space exploration, and I thank my students for thoughtful discussion.

I am also thankful for the support and feedback from colleagues not yet mentioned that I have met either at conferences or through correspondence. In no particular order, these individuals include: Les Johnson, Linda Billings, Laura Delgado Lopez, Sean McMahon, Erik Persson, John Rummel, Margaret Race, Mark Lupisella, Stephen Baxter, Janet de Vigne, Mukesh Bhatt, Luke Matthews, Ian Crawford, Kathryn Denning, David Hewitt, Michael Turner, Dan Lester, Eleni Panagiotarakou, Christopher Newman, Robert Kennedy III, Cameron Smith, Seth Baum, Annalea Beattie, Arne Lahcen, John Lewis, Doug Loss, Zach Pirtle, Sheri Wells-Jensen, Keith Abney, Colin McInnes, Oz Monroe, Joseph Gottlieb, William Kramer, and anyone else I have forgotten to mention.

Finally, I would like to thank my editor at Oxford University Press, Sarah Humphreville, for supporting this project since the moment she received my first inquiry. Without her encouragement this project would not have been possible.

Chapter 1 summarizes my "Myth-Free Space Advocacy" series of papers (Schwartz 2017a; 2017b; 2018a; 2020) in which I highlight the lack of evidence supporting many common spaceflight rationales. However, my evaluation of spaceflight's impact on STEM degree conferral rates is considerably more comprehensive and rigorous than the assessment I provided in Part III of this series (Schwartz 2018a). Chapter 3 includes a substantially reworked and expanded version of an argument for the state support of science that I first made in the context of space societies (Schwartz 2015b). Chapter 4 works with ideas about the rationale for and scope of planetary protection originally developed in Schwartz (2019b). In my discussion of space resource exploitation and regulation in Chapter 5 I have substantially overhauled and updated arguments, concerns, statistics, and figures that I presented originally in Schwartz (2014) and Schwartz (2016b) and in a joint paper with Tony Milligan (2017). The basic position I take in Chapter 6—that we ought to prioritize space science over space settlement—is articulated very briefly in Schwartz (2019a) and Schwartz (2019c). Finally, my discussion of obligations to space settlers borrows ideas that I first developed in the context of thinking about multi-generational worldship travel (Schwartz 2018b). I thank Elsevier, Wiley and Sons, Springer, and Taylor and Francis for permission to reprint various excerpts and figures.

—JSJS, June 2019

Introduction

Subtle Shift in Emphasis

For the majority of the past decade, *commercial* spaceflight has dominated discussion about space exploration. Most likely to be featured in popular science and technology news are, e.g., the efforts of Virgin Galactic to open up the suborbital tourism industry and the ambitions of Space Exploration Technologies Corporation (SpaceX) to settle Mars, whereas current *scientific* exploration initiatives, for example the Mars *InSight* mission of the National Aeronautics and Space Administration (NASA) or the *ExoMars* mission of the European Space Agency (ESA), receive comparably less attention. This is perfectly understandable, of course, as the leaders of Virgin Galactic and SpaceX—Richard Branson and Elon Musk, respectively—are well-known and outspoken public figures. Scientific missions, meanwhile, take a considerable amount of time, and, with the exception of the momentary excitement of mission launches (and landings, in some cases), they do not attract terribly much public interest. (Granted, considerable public attention arises when things go wrong, as in Apollo 1, Apollo 13, STS-51-L (the *Challenger* disaster), STS-107 (the *Columbia* disaster), or, less dramatically, the failure of the *Schiaparelli* lander.) Moreover, the United States and Luxembourg have passed legislation encouraging commercial spaceflight. The US Commercial Space Launch Competitiveness Act of 2015 not only encourages NASA to rely more heavily on the private sector for launch services, it also encourages private industry to develop capabilities associated with space resource exploitation, such as lunar and asteroid mining.

There are many reasons for the growing utilization of the private sector for the provision of services to and in space, but far and away the most significant factor is cost. It is incredibly expensive to launch material into low-Earth orbit (LEO), with launch costs in the range of 2,000–10,000+ USD/kg (and with even greater per-kg expenses associated with missions to more energetically distant destinations, such as geostationary orbit (GEO), the Moon, or the other planets or their satellites). The hope, then, is that by encouraging greater private sector participation and competition in the design, construction, and use of launch vehicles and spacecraft, space missions will become more affordable. However, this brings to the fore questions about what the fundamental purpose of spaceflight is or should be. The recent focus on commercial spaceflight suggests that, deep down, spaceflight exists to provide new markets for economic activity.

Despite this shift of attention from national and international space programs to private initiatives, the rhetoric surrounding space exploration has changed little since the *Apollo* program of the 1960s and 1970s. Whether one is advocating for increasing the budget of NASA or ESA, for withdrawing from the United Nations Outer Space Treaty and its restrictions on commercial exploitation of space resources, or for fast-tracking SpaceX's plans to settle Mars, it is very easy to predict the kinds of reasons that will be raised: that we need to explore space to save humanity from extinction; that we need to use space resources to prevent further destruction of Earth's ecosystems; that we need to explore space to inspire student interest in the sciences; that we need to conquer the space frontier to avoid societal stagnation; that humans are explorers; etc. Reasons like these tend to come together as something of a "space advocacy package." However, the argumentation that is usually supplied in support of these claims (if any is supplied at all) tends to be severely lacking in quality and rigor. It is almost as though it is a tradition, or a tenet of some kind of spaceflight religion, that one must speak these "spaceflight truths" whenever one defends spaceflight.

The problem is that advocates seldom bother assembling evidence in support of the various rationales they provide for spaceflight. Rather, space advocacy consists mostly in the repetition *ad nauseum* of a small set of talking points, possibly with an appeal to the authority of astronauts, astrophysicists, or world leaders. But if you want to know whether spaceflight is educationally inspiring, or whether humans have an innate tendency to explore, astronauts and astrophysicists are not the correct authorities to seek out. You would do better to consult with education scientists, sociologists, psychologists, and evolutionary biologists. There is a minimum attention to *appropriate* sources of evidence that is lacking in much of space advocacy. And if you are someone such as myself who believes that spaceflight is incredibly important, but also that its importance should be established through good reasoning, then you should be far from satisfied with the current state of space advocacy.

Philosophers by training are especially attentive to the contours of reasoning. That is, they tend to focus, and in great detail, on the formulation, discernment, and evaluation of arguments. One result of this is a tendency for philosophers to adopt unusually high standards when it comes to accepting and rejecting beliefs. So you might think it is only because of disciplinary bias that, as a philosopher, I tend to be skeptical of basic spaceflight rationales. And though I intend for this book to contribute to professional philosophical discourse, nevertheless most of what I have to say will be accessible, meaningful, and relevant to individuals from a wide range of disciplinary and vocational backgrounds—from planetary scientists to political scientists; from astrobiologists to anthropologists; from space program employees to lawyers and legal scholars. In other words, the reasons why we should reject most basic tenets of space advocacy, and the reasons I shall offer in their place, should be persuasive to more than just my fellow philosophers. Indeed, they should be compelling to anyone with an interest in space exploration that also shares an interest in reason- and evidence-based belief formation. With any luck, what I have to say will also be compelling to many who are inclined to doubt the value of spaceflight. But regardless of whether my own conclusions are ultimately correct, I will consider this book a success if it motivates readers to increase the care with which they reason about spaceflight and its importance.

What reasons, then, do I plan to offer in support of spaceflight? I approach this as a question within ethics, broadly speaking. My focal position is that something especially good would be realized or facilitated by spaceflight, *viz.*, *the generation of scientific knowledge and understanding*. Speaking roughly, by "scientific knowledge" I mean a true belief in some domain of science that is justified based on the best evidential practices of that domain; by "scientific understanding" I mean systematic knowledge of some topic or theory in some domain of science along with the ability to deploy this knowledge in appropriate situations. Consequently, what I shall argue for is that spaceflight provides us with a vital and productive avenue for increasing our knowledge and understanding of ourselves, our planet, our Solar System, and our Universe. Scientific knowledge and understanding are not only intrinsically valuable (i.e., valuable and worth seeking for their own sakes); they are also instrumentally valuable (i.e., valuable on account of their good consequences or effects) for the ways they contribute to overall societal welfare and progress. Neither of these points is inherently consequential; in place of talking about the importance of *having* and *using* scientific knowledge and understanding we could instead talk about the importance of, e.g., *being the kinds of persons* who seek scientific knowledge and understanding. (I hope to stay neutral, as far as can be maintained, when it comes to theoretical disputes arising in normative ethics regarding the ultimate nature of right or wrong action.)

My proposal raises several questions. First, "spaceflight" is an unhelpfully general term. It includes activities such as crewed and robotic exploration of the space environment, human space settlement, suborbital space tourism, space resource exploitation, Earth observation from space, military and commercial satellites services, etc. When I argue that spaceflight should be supported because it contributes to the generation of scientific knowledge and understanding, then, do I intend to support all of these activities equally, or to support only some of them? My answer is that this support is restricted to just those activities most likely to make meaningful contributions to science. Thus, I will not argue for spaceflight universally but specifically for activities such as Earth observation, as well as robotic and crewed scientific exploration missions. In contrast, commercial and military spaceflight activities are capable of accomplishing far less, ethically speaking. It is therefore concerning that the public seems much more interested in, e.g., SpaceX and in the possible creation of a new branch of the US military devoted to space security, than they are in all of the incredible work being done by space scientists at universities and space agencies around the globe.

More substantively, my view has consequences for what activities ought to be prioritized when it comes to the exploration and use of space. As I shall argue in various places throughout this book, there is a significant risk that scientific and non-scientific uses of the space environment will come into conflict. Since greater goods are associated with scientific uses of the space environment, these should be preferred whenever they conflict with other proposed uses of space. That is, scientific objectives ought to be pursued in advance of, or in place of, commercial objectives. If, for instance, we are faced with an exclusive choice between sending a mission to an asteroid in order to mine this asteroid for metals or other resources, and sending a mission to an asteroid in order to study its composition and to learn about the early history and evolution of the Solar System, we ought to select the latter.

Importantly, my prioritization of scientific uses of the space environment is limited in its time-horizon and rests on (what some might regard as) conservative assumptions about the progress of spaceflight technology. My focus throughout will be spaceflight of the present, as well as the "near future" (which I take to extend two centuries into the future). Within this time frame, I assume that no game-changing spaceflight technologies will arise. That is, I shall assume that we will not acquire technologies belonging more the realm of science fiction (warp drive; wormhole travel); that the frequency of space launches (and crew complements) will not increase by much more than one or two orders of magnitude; and that interplanetary transit times will remain more or less constant.[1] I cannot guarantee that I would defend an identical set of conclusions outside of these constraints, although I explore this issue very tentatively in the Epilogue.

This brings us to the second question that my position raises: Is it the case that scientific knowledge and understanding are valuable in the ways I have

---

1. Of these assumptions, the last is the least safe: It is not unreasonable to suppose that either nuclear, antimatter, or directed energy propulsion systems will be implemented within the next two centuries, which could dramatically decrease interplanetary transit times.

suggested? That is, are scientific knowledge and understanding worth pursuing for their own sakes? And would their pursuit be likely to realize various further societal goods? I suspect that most readers of this book probably already agree that scientific knowledge and understanding are valuable in these ways. However, this is another area in which little support tends to be offered by space advocates (as well as by science advocates and philosophers more generally). So, while intuitively it is the case that scientific knowledge and understanding are valuable, intrinsically as well as instrumentally, I would prefer to demonstrate rather than assume their value. These are not simple tasks, as I shall explain momentarily.

A third question is: How do the values associated with using spaceflight to generate scientific knowledge and understanding compare with the values associated with using spaceflight for other purposes? As I mentioned a moment ago, my intention is to show that scientific spaceflight should be preferred to non-scientific spaceflight. But that seems difficult to maintain because weighty obligations other than those related to science support non-scientific spaceflight. One of our strongest obligations is to ensure the long-term survival of the human species. Since humans cannot survive on Earth forever, we must attempt to create permanent, self-sustaining human societies in space. Another strong obligation is to provide for and improve the material well-being of humanity. Since Earth's resources are limited, we must look to space to support the resource needs of the human species. An obligation to use space to generate scientific knowledge and understanding, while laudable, detracts from securing the much greater goods of ensuring human survival and well-being.

The attractiveness of my view, then, depends on two things: First, that the argument for space science is *stronger* than traditionally assumed. And second, that the arguments for other forms of spaceflight are *weaker* than traditionally assumed. Free of any particular context, I would agree that human survival and well-being matter more than scientific knowledge and understanding. Thus, I shall not defend the claim that obligations to ensure human survival and to provide for human well-being are weaker than an obligation to pursue scientific knowledge and understanding. But the strength of our obligations *in practice* is very much a contextual matter. As a corollary to the idea that "ought implies can," if we lack any effective, affordable means for satisfying an obligation—even a very strong obligation—then in practice this obligation is neither strong nor overriding. Meanwhile, if we possess an effective, affordable means for satisfying an obligation—even a very weak one—then in practice the strength of this obligation increases. Over the near future, if it can be shown that spaceflight either fails altogether to ensure human survival, or neither effectively nor affordably ensures human survival, then for the moment there is no strong or overriding duty to use spaceflight for this purpose. At the same time, if it can be shown that spaceflight is an effective and affordable means for generating scientific knowledge and understanding, then we would have a comparatively stronger duty to use spaceflight for this purpose. It is through this kind of reasoning that I intend to establish the relative priority of the value of space science.

I begin this project in Chapter 1, in which I consider and reject many of the standard space advocacy arguments. First is the argument that spaceflight is educationally inspiring, i.e., that money spent on spaceflight increases student interest in science, technology, engineering, and mathematics (STEM) disciplines. There are, unfortunately, few clear positive connections between undergraduate and graduate degrees in STEM fields and spaceflight funding. So, we lack the kind of statistical evidence that would be the first step in building a causal argument that spaceflight has a substantive impact on students' educational choices. Second is the argument that spaceflight promises to reveal answers to questions of universal significance, for instance about the origin of human life, and about whether there exists extraterrestrial life. Though relatively little survey data exist on these topics, the data available suggest that most people are not especially interested in what science has to say about the origin and extent of life.

A third of the standard rationales is that humans have a basic urge to explore which justifies human space exploration and settlement. As genetic and anthropological research has revealed, there are several genes associated with exploratory behavior (which refers primarily to activities like local reconnaissance) and past human migration. But these associations do not provide evidence of a basic urge to explore, since research also indicates that traits like exploratory behavior did not *impel* past migrations but were instead selected for *subsequent* to them. At present, there is no known genetic or biological basis for claiming that humans have an innate tendency to see what is over the horizon, no less to branch out into the space environment. A fourth rationale is that human society needs a new, space frontier to avoid stagnating. In conquering the Martian frontier, settlers would be confronted with a challenging environment. This would force them to improvise, innovate, and adapt—all in ways that will teach the rest of humanity valuable lessons about science, technology, and democratic governance—just as the conquering of the American West did for the United States. In addition to being historically dubious, this strain of thinking dramatically underestimates the potential for space settlement to provide unwelcome lessons. For instance, in order to survive in the instantaneously lethal conditions on Mars, settlers might adopt autocratic or totalitarian forms of governance. The result might be an exercise in human suffering, as opposed to the energizing of democratic culture.

The tenebrous quality of these rationales is the starting point for the positive project of the book, which is to provide an articulation and defense of the value of space science. The first (and most philosophically technical) task here is to argue for the intrinsic value of scientific knowledge and understanding, which I undertake in Chapter 2. Here I parley with several debates in contemporary epistemology regarding the value of knowledge and the value of understanding. What emerges from this is that the value of knowledge comes primarily from the value of true belief; while the value of understanding comes from the value of true belief as well as the value of cognitive achievement. The central task here is to provide a defense of the intrinsic value both of true belief and of cognitive achievement. I argue that each value can be demonstrated on a broadly naturalistic approach to justifying attributions of intrinsic value, according to which an entity is intrinsically valuable whenever its being valued for its own sake functions as part of the best explanation of some practice implicated by the scientific worldview. The valuing of true beliefs and cognitive achievements for their own sakes passes this test. Therefore, true beliefs and cognitive achievements are intrinsically valuable—especially those related to science generally, and space science in particular.

Chapter 3 takes up the task of defending the instrumental value of scientific knowledge and understanding. The underlying argument here is that increases in scientific knowledge and understanding contribute to social progress. Furthermore, increases in scientific knowledge and understanding are spurred in large part by scientific exploration and research. This is because scientific exploration is vital for the collection of new observations as well as the testing of our existing scientific ideas, concepts, and theories. Space exploration is implicated because it is a form of scientific exploration that is especially likely to contribute to scientific progress. Also discussed in Chapter 3 is the strength of the obligation of democratic states to provide support to scientific research. Employing recent work in social epistemology and political science, I argue that democratic states have a duty to support scientific research when that research promises to contribute to the democratic process in certain, important ways. Using numerous examples, I show that space science makes vital contributions to the democratic process and therefore ought to be supported by democratic states.

Chapters 2 and 3, then, are intended to provide a positive account of the value of space science. The remainder of the book applies these values to several key issues in contemporary space exploration and policy. In Chapter 4 I discuss the role of the value of science in debates about the rationale and scope of planetary protection policies. Biological and other contamination of the space environment present unique challenges. Unless we can be confident that Mars or other locations have not been contaminated by terrestrial organisms, we could not rule out that any potential discovery of extraterrestrial life might be a supremely expensive false positive. For this reason, space programs implement various policies to minimize the risk of contaminating sites of interest in the search for extraterrestrial life.

Only in the past three decades, however, has it been recognized that anything of ethical significance might hang in the balance. Perhaps planetary protection is needed, not only to protect space environments for the sake of the scientific search for life, but also *for the sake of any endemic life* we might find in these environments, even if that life is only microbial in nature. Arguments have been put forward to the effect that any extraterrestrial life that we find would be intrinsically valuable, and therefore, the protection of this value is the ideal purpose of planetary protection. While I shall respond to some of the criticisms that have been levied against these arguments, nevertheless my own position is that preoccupation with protecting extraterrestrial life for its own sake needlessly limits the scope of what should be said about the ethics of planetary protection.

Other values that must be protected in the space environment include the intrinsic and instrumental values of the knowledge and understanding that can be generated via the scientific investigation of the space environment. We have an obligation to conserve opportunities for carrying out scientific exploration and research that facilitate increases in our knowledge and understanding of the space environment. This points to a much broader obligation to protect the space environment against contamination or disruption, as much more of science is at stake than astrobiology and the search for extraterrestrial life. The space sciences *in toto* have yet to disclaim interest in any space environment, planet, moon, or other celestial body. Since scientific exploration is most effective in pristine environments, then as a precautionary default we ought to assume that space environments are of interest to science until proven otherwise.

In Chapters 5 and 6 I discuss how my approach to the value of space science and to planetary protection figures in an assessment of two issues currently receiving considerable attention in public discussions about space exploration: the exploitation of space resources, and space settlement. In Chapter 5 I argue that space science objectives ought to be prioritized over the commercial exploitation of space and its resources, and therefore that we should deny attempts to revise or otherwise supplant the Outer Space Treaty, which is thought to be prohibitive of commercial space resource exploitation. Those pushing for regulatory relief typically justify space resource exploitation by pointing to its purported ability to free humanity from the various harms associated with terrestrial pollution and resource depletion. By using space resources, for instance from the Moon and from near-Earth asteroids, the hope is that we can provide humanity with additional raw materials and energy while also moving polluting industries into space. Against this I argue that there are severe restrictions on *accessible* space resources. The resources of space, while staggeringly vast in principle, are nonetheless non-renewable and rather limited in practice. For instance, based on recent research the amount of water that could be melted from lunar polar ice, together with all of the water that could be mined from the asteroids which are as energetically accessible as the Moon, only comes to about 3.7 km^3 in volume. This does not support conceiving of space resources as limitless or as capable of freeing us from the need to more diligently combat pollution and resource depletion on Earth. While we have a strong collective obligation to limit terrestrial pollution and to mitigate the effects of terrestrial resource depletion, we cannot satisfy this obligation effectively via space resource exploitation. Therefore, this is a use of the space environment that we ought for the moment pass over in favor of our obligation to preserve space for scientific investigation.

A structurally similar defense of space science is also available against space settlement. As I discuss in Chapter 6, there is a compelling obligation underlying the apparent need to pursue permanent, self-sustaining space settlements, *viz.*, the obligation to ensure the long-term survival of the human species—or, as I shall follow Tony Milligan in calling it, the duty to extend human life. Here there are two arguments that need to be distinguished: an in-principle argument according to which space settlement is *ultimately* necessary for extending human life; and a pressing argument according to which space settlement is *urgently* necessary for extending human life. I accept the conclusion of the in-principle argument, and furthermore I defend it against several objections, including some that cast doubt on the existence of an obligation to extend human life. However, I shall argue that space settlement is not needed urgently (i.e., in the near future) because most of the major threats to human existence (asteroid collisions, ecological collapse, etc.) can be addressed more effectively by other means. For instance, if what you want to do is minimize the risk of human extinction via asteroid impact, the best thing to do would be to protect Earth by increasing funding for asteroid detection and diversion initiatives. This would not only do a better job of lowering the risk of human extinction; it would also be much cheaper. Since there is no pressing or immediate need to settle space, then we can for the time being preserve the space environment for scientific study.

Over the long term, however, space settlements will be needed, which raises the question as to whether it would be permissible to subject future generations of space-dwellers to the conditions of life in a space settlement. Life in space will typically consist of living within artificial habitats (which would be necessary to protect settlers from the intensely hostile environments found throughout the Solar System). In contrast with living on Earth, living in a space habitat would be incredibly confining, both physically and emotionally, and afford settlers with little privacy, and with little freedom of choice when it comes to education, vocation, and romantic companionship. As I argue in the second half of Chapter 6, this unveils space settlement as potentially unjustifiably exploitative of persons born into the settlement, who may have no choice except to live under unacceptable conditions. Therefore, a requirement on ethically permissible space settlement is that the settlers can provide reasonable guarantees that their children will not experience unduly exploitative conditions.

In the Epilogue, I provide a re-emphasis of the importance of science and a brief discussion of how relaxing certain assumptions (about time-horizons and technical capabilities) might affect our obligations in space. What emerges here, but also throughout the book, is the enduring importance of scientific research—not just to contemporary societies but also to any societies that might develop in the space environment. Beyond their intrinsic value, scientific knowledge and understanding remain of great instrumental value to anyone interested in creating and sustaining human life in space. Indeed, the democratic case for the support of scientific research articulated in Chapter 3 applies even more strongly in the case of space societies, which will be even more dependent upon research for their survival and flourishing.

It follows, then, that scientific knowledge and understanding, including that which is generated by space exploration, will remain of significant value to human society. Science therefore is and should remain the predominant stakeholder when it comes to decisions about spaceflight funding priorities, spaceflight mission objectives, and legislative and other policy initiatives. There is great potential for commercial spaceflight to help lower launch costs and increase payload capacities. But it ought to remain a handmaiden to the space sciences, rather than encouraged to become an invasive species which competes with space science for resources. I hope, therefore, that this book provides a compelling, illuminating, and philosophically satisfying basis for recapturing the attention that space science appears to have lost to the "New Space" movement, but that it clearly deserves.

CHAPTER 1

Rationales for Space Exploration

A Review and Reconsideration

Space exploration is expensive and full of risk. Even with recent cost reductions stemming from increased competition among launch vehicle manufacturers, reliable payload delivery to low-Earth orbit (LEO) remains well in excess of 2,000 USD/kg, with even greater costs associated with higher orbits (such as MEO (medium-Earth orbit) or GEO (geostationary orbit)), or with interplanetary missions. Human missions to LEO cost on the order of tens of millions of USD per passenger. Meanwhile, national spaceflight budgets tend to be quite small. In 2017, the budget of the National Aeronautics and Space Administration (NASA) was approximately 19.5 billion USD—0.47% of U.S. federal outlays. In the same year, the budget of ESA was 7.1 billion in USD (5.8 billion euro); the budget of JAXA (Japan Aerospace Exploration Agency) was 1.4 billion in USD (154 billion ¥); the budget of ISRO (Indian Space Research Organization) was 1.2 billion in USD (8045 Crores); and in 2015 the budget of Roscosmos (Russia's space agency) was 2.9 billion in USD (186.5 billion rubles). What's more, even the most reliable launch systems, such as ESA's *Ariane 5*, the United Launch Alliance's *Atlas V*, etc., have failure rates of 1% to 5% or higher.[1] Human-rated systems are no exception; 2 out of 135 NASA Space Transportation System missions resulted in catastrophic failure—the *Challenger* (STS-51L) and *Columbia* (STS-107) disasters of 1986 and 2003. In 2014, a confluence of issues, including pilot error, led to

---

1. See Federal Aviation Administration, *The Annual Compendium of Commercial Space Transportation: 2018*, available at: https://www.faa.gov/about/office_org/headquarters_offices/ast/media/2018_AST_Compendium.pdf (accessed 4 June 2019).

the death of one of the pilots of Virgin Galactic's *VSS Enterprise*. Two Soviet missions suffered fatalities—*Soyuz* 1 in 1967 (because of a parachute failure during landing), and *Soyuz 11* in 1971 (due to spacecraft decompression; the only case to date of fatalities in space). It should come as little surprise, then, that spaceflight advocates have felt the need to engage in lively defenses of continued or expanded spaceflight activities.

There is something like a "spaceflight advocacy package" of arguments that tend to be employed whenever the topic is broached. Among the battery of spaceflight rationales often promulgated are:

- Increases in spaceflight activities will inspire more students to become interested in science, technology, engineering, and mathematics (STEM) disciplines.
- Through the exploration of the Solar System, and especially through the search for extraterrestrial life, space exploration promises to deliver answers to many of life's "big questions" about life's extent and origin.
- Crewed spaceflight is a natural expression of our innate exploratory and migratory urges.
- The "conquering" of the "space frontier" will realize various societal goods purportedly realized by the "conquering" of the American West.
- The exploitation of resources from space (e.g., from lunar and asteroid mining) will promote human well-being by mitigating terrestrial resource depletion.
- Human expansion into space is a necessary means for preserving humanity against global terrestrial catastrophes (e.g., ecological collapse, meteorite strikes).
- Space exploration is a critical driver of technological innovation (the "spinoff" justification).

As this list evinces, the term "space exploration" is ambiguous.[2] After all, a supporter of "space exploration" could be a supporter of: the scientific study of space environments (either crewed or robotic); the use of space for commercial purposes (e.g., space hotels, space mining); the use of space for human habitation; etc. The significance of this ambiguity is that there is no such thing as a rationale for space exploration *simpliciter*. Rather, there are many and varied rationales for the many and varied possible *objectives* or *activities* that might be undertaken in space.

A central question of this book is whether and in what senses we are justified in saying there is a *moral obligation* to support space exploration. A necessary condition on the existence of an obligation to satisfy a spaceflight

2. A point first brought to my attention by Lester and Robinson (2009).

objective is that, *ceteris paribus*, some amount of good would come from satis-fying this objective. However, this condition is insufficient on its own, because other things may not be equal. It may be that the opportunity costs are too high, i.e., that we could realize more good by working toward other objectives. Or it might simply not be possible for the moment to satisfy some spaceflight objective. For example, it may be true in principle that it is good to mitigate terrestrial resource depletion by exploiting space resources. Similarly, it may be true in principle that it is good to ensure long-term human survival via space settlement. But in each case there are good reasons to doubt that we can, at present, accomplish these tasks effectively.[3] This motivates two further necessary conditions on the existence of an obligation to satisfy a spaceflight objective: that it is possible in the first place to do this; and that doing this is a justifiable use of energy and resources. If some spaceflight objective satisfies all three conditions, then we know that, in addition to being good to realize in principle, it is also a good that it is possible for us to realize, and one that can be defended as worthwhile against other possible uses of our energies and resources.

Importantly, whether and to what degree a spaceflight objective satisfies the three conditions—of being good in principle, of being possible to realize, and being good on balance—varies over time. At any given time, certain spaceflight objectives might offer more good (in principle) than others, and *ceteris paribus*, we have an obligation to prioritize those objectives most likely to bring about the most good.[4] At present, inspiring students to study STEM disciplines may be more important, all things considered, than establishing permanent space settlements. But several centuries into the future it could well be that space settlement becomes a prominent societal goal. Similarly, the *possibility* of satisfying spaceflight objectives varies over time. As an ap-plication of "ought implies can," if a particular spaceflight objective is beyond our scientific, technological, or economic capabilities, then it could not be demanded that we satisfy this objective.[5] Thus, if it is presently within our power to use spaceflight to inspire students to study STEM disciplines, but not in our power to establish space settlements, then our present obligations attach more strongly to the former objective than they do to the latter.

Furthermore, our ability to satisfy spaceflight objectives *effectively* also varies. Perhaps it is possible that we can use spaceflight *both* to increase STEM enrollments and to establish extraterrestrial settlements. Still, the former would arguably be much easier, much less expensive, and much less risky. For

3. I will outline my reasoning on these two topics later in this chapter, though the details will have to wait until Chapters 5 and 6, where I discuss space resource exploi-tation and space settlement, respectively.

4. Which is not to say that the attainment of good is always instantaneous.

5. We might however bear an obligation to attempt to build the capacity to satisfy the objective.

this reason, we might judge that for the moment we have a stronger obligation, all things considered, to use spaceflight to increase STEM enrollments. (And with respect to goals such as increasing STEM enrollments, there is always the possibility that some alternative means is more efficient and effective!) We must keep in mind, then, that we have numerous obligations, some of which come into conflict, and that many of these obligations admit of multiple means of satisfaction, some of which conflict. This means that the existence and strength of an obligation to engage in any particular activity is a highly contextual matter that cannot simply be read off of the mere observation of an apparent need together with an apparent means for addressing this need. What we in fact have an obligation to work toward in the present might look very different from what we are obligated to do decades and centuries into the future.

I will happily grant that virtually every spaceflight objective satisfies the first condition—that, *ceteris paribus*, good would be realized by increasing STEM enrollments; searching for evidence of extraterrestrial life; satisfying desires to explore; etc. It is perfectly coherent that, *in principle*, we have obligations to satisfy a wide variety of spaceflight objectives. Moreover, these obligations exist at the collective level. An obligation to increase STEM enrollments is borne not by any persons in particular, but by humans *collectively*, perhaps through institutions with the requisite resources and capacities. Such an obligation also exists *for the sake of* humans collectively as opposed to existing for the sake of any particular human. Likewise for the obligation to search for extraterrestrial life, etc.[6] The goal of this book, however, is not to produce a complete list of *prima facie* duties related to spaceflight, but rather to determine which uses of space are most important both at present and in the near future (a period I take to extend roughly two centuries into the future), i.e., over timescales where we have considerable knowledge about our collective needs, our social, political, and technological capabilities, and the scope and extent of our other obligations. And in this context other things are far from equal. Many proposed spaceflight objectives do not satisfy the necessary condition of being possible to accomplish in the present or in the near-future. And many proposed spaceflight activities are not especially effective means for satisfying their corresponding obligations, and thus fail to satisfy the necessary condition of being an all-things-considered justifiable use of energy and resources.

With this in mind I would like to provide an assessment of traditional spaceflight rationales. As I shall argue, nearly all of the rationales mentioned at the outset fail to justify corresponding obligations in the present or near future. The only obligations which remain largely unscathed are those related to the scientific study of space environments. It will be the job of the rest of the

---

6. I will not discuss whether such obligations exist apart from, supervene on, or are reducible to more characteristically individualized duties.

book to argue that the scientific exploration of space is particularly valuable, and that it should be prioritized over other spaceflight objectives.

EDUCATIONAL INSPIRATION

A common theme of space advocacy—both in the popular science literature as well as in the peer-reviewed science and space policy literature—is that spaceflight ought to be given increased support because it is uniquely educationally inspiring. Such inspiration could be evidenced either through increased interest in STEM disciplines (as manifested by STEM enrollments) or though increases in scientific literacy among the general public. I shall grant for the duration that it would be a good thing to realize both of these goals. I am particularly uninterested in questioning the value of increasing the scientific literacy of the general public; however, I grant that it is contentious that society (at least, American society) is currently in need of more individuals with STEM degrees, as enrollments in engineering and computer science in particular have shown strong growth over the last 30 years (but perhaps the complaint is that they have not grown quickly enough?). Nor will I argue against the possibility of accomplishing either of these tasks. What I will argue is that there is no clear evidence that spaceflight is an effective generator of either kind of inspiration. Since supporting spaceflight is not an effective way of satisfying the obligation (if there be one) to inspire interest in science, we have no associated obligation to support spaceflight.

Space and STEM

Consider first the claim that spaceflight activities are critical drivers of STEM degree conferrals. One proponent here is Eligar Sadeh, who claims there is evidence of an apparent connection between NASA's budget and the rate at which students in the United States earn degrees in STEM disciplines:

> During Apollo, there was a dramatic increase in the number of U.S.-citizen students pursuing advanced degrees in STEM disciplines. As the Apollo program was terminated and NASA's funding cut, the number of students going into STEM fields correlated with the downward trend in NASA's budget, especially with regard to graduate studies at the Ph.D. level . . . .
> A contributing factor to these trends is a general disinterest in STEM fields. The argument that money put into the space program is better spent by putting it directly into the educational system to encourage students to pursue STEM areas is a misconception, as the United States is already one of the top spenders per student in the world. The bottom line is that students need something to

inspire their efforts. Thus, the positive impact of space exploration on STEM education is without precedent, as is evident with aspects of the Apollo paradigm and the inspirational value of Apollo. (2015, 531–532)

Sadeh is correct that there is a striking correlation between NASA's budget and degree conferral rates in STEM disciplines during the Apollo era (roughly, the period from 1960 to 1975), at least when measuring the former in absolute dollars and the latter in total doctoral degrees conferred in various STEM disciplines. But as Table A.52 in Appendix A shows, during the same period doctoral conferral rates in most disciplines also positively correlated with NASA's budget, and in many cases even more strongly with the total federal budget. Thus it would be overhasty to identify NASA's activities as the primary drivers of degree production, especially when it is noted that, over the same period, overall federal spending on science exhibited roughly the same pattern—a spike in the mid-1960s, followed by a steady return to prior levels of funding (at least as a percentage of the federal budget) by the mid- to late-1970s. A more careful hypothesis would be that students were responding to—i.e., "inspired by"—perceived increases in employment opportunities in various sectors, including the space sector.

A provisional assessment of this hypothesis is possible by comparing overall U.S. degree conferral rates (bachelor's, master's, and doctoral degrees) with various categories of U.S. federal outlays.[7] At the time of analysis (the fall of 2018), comprehensive data on degree conferral rates was available only for the period between 1970 and 2015—the years for which data was available for all disciplines from the National Center for Education Statistics' Digest of Education Statistics.[8] For this reason the period from 1970 and 2015 will be my focus.

Multiple linear regression models were used to address whether variation in degree conferral rates can be explained by variation in funding levels in *a priori* relevant areas. These models were constructed to address *targeted* questions such as "Does funding for biomedical research predict biomedical degree conferral rates?" as opposed to *general* questions such as "What, if anything, predicts biomedical degree conferral rates?" In order to control for the influence of population growth, fluctuations in annual budgets, etc., degree conferral rates were input as percentages of the U.S. population and funding levels were input as percentages of total federal outlays. For each discipline, three models were constructed factoring for three different delays between

7. Thanks to Wendy Whitman Cobb for helpful discussion on the educational impact of federal funding.

8. The Survey of Earned Doctorates contains accessible data dating back to the 1920s, but uses different disciplinary categorizations, and obviously does not provide information on either master's or bachelor's degree conferral rates.

funding year and degree conferral year—four years, six years, and eight years. Thus, degrees conferred in 1970 are associated with funding levels in 1966, 1964, and 1962, respectively; degrees conferred in 1971 are associated with funding levels in 1967, 1965, and 1963; etc. The justification for this is that changes in funding are unlikely to have an immediate impact on degree conferral rates, but may be more likely to impact conferral rates some years later. I will only discuss the results of the analyses, but see Appendix A for methods, results, and data sources. Only positive associations significant at $p < 0.05$ are mentioned here.

Disciplines for which funding sources can explain the majority of variation (adjusted $R^2 > .500$) in degree conferral rates include:

- Agriculture degrees, which on a four-year delay are positively associated with energy and natural resources funding. On six- and eight-year delays degrees are positively associated with agricultural research. See Tables A.1–A.3.
- Biology and Biomedical Research degrees, which for all delays are positively associated with funding for healthcare services, health research and training, and NASA. By an order of magnitude the strongest contributor is funding for health research and training (NASA funding is only significant at $p < 0.05$ on a four-year delay). See Tables A.7–A.9.
- Communications degrees, which on a four-year delay are positively associated with funding for health care services and labor services, with the latter being the stronger contributor by two orders of magnitude. On six- and eight-year delays the only significant positive association was with funding for health care services. See Tables A.13–A.15.
- Computer Sciences degrees, which on a four-year delay are positively associated with funding for NASA. On six- and eight-year delays the only significant positive association was with funding for the National Institutes of Health (NIH). See Tables A.16–A.18.
- Engineering degrees, which on a four-year delay are positively associated with funding for military research and development, energy, Health and Human Services (HHS), and the Corps of Engineers/Civil Works. By an order of magnitude the strongest contributor is funding for the Corps of Engineers/Civil Works. On a six-year delay engineering degrees are positively associated with funding for military research and development, energy, HHS, the Corps of Engineers/Civil Works, and the Environmental Protection Agency (EPA)—with the strongest contributor again being funding for the Corps of Engineers/Civil Works. On an eight-year delay degrees are positively associated with funding for military research and development, HHS, and the EPA—with funding for the EPA being the strongest contributor by an order of magnitude. See Tables A.22–A.24.

- English and Literature degrees, which on six- and eight-year delays are positively associated with funding for higher education and elementary, secondary, and vocational education. See Tables A.25–A.27.
- Foreign Languages degrees, which on six- and eight-year delays are positively associated with funding for higher education and elementary, secondary, and vocational education. See Tables A.28–A.30.
- Health Professions degrees, which for all delays are positively associated with funding for health care services and health research and training. On an eight-year delay the latter is a stronger contributor by an order of magnitude. See Tables A.31–A.33.
- Mathematics and Statistics degrees, which on a four-year delay are positively associated with funding for military research and development. On a six-year delay degrees are positively associated with funding for NASA. On an eight-year delay degrees are positively associated with funding for the Department of Energy and for the National Science Foundation (NSF), with the latter being stronger by an order of magnitude. See Tables A.34–A.36.
- Physical Sciences degrees, which on a four-year delay are positively associated with funding for energy, NASA, and the NSF. On a six-year delay degrees are positively associated with funding for military research and development, energy, and the NSF. On an eight-year delay degrees are positively associated with funding for military research and development, natural resources, and the NSF. In each case funding for the NSF was the strongest contributor by an order of magnitude. See Tables A.37–A.39.
- Psychology degrees, which on four- and six-year delays are positively associated with funding for health care services and health research and training, with health research and training being the stronger contributor by an order of magnitude. On an eight-year delay degrees are positively associated with funding for health care services. See Tables A.40–A.42.
- Public Administration and Social Work degrees, which on a four-year delay are positively associated with funding for health care services and for consumer and occupational health and safety, the latter being the stronger contributor by three orders of magnitude. On a six-year delay degrees are positively associated with funding for health care services. On an eight-year delay degrees are positively associated with funding for health care services and the Department of Education; the latter being the stronger contributor by an order of magnitude. See Tables A.43–A.45.
- Social Sciences and History degrees, which on a four-year delay are positively associated with funding for the NSF. On a six-year delay degrees are positively associated with funding for labor services and the NSF. On an eight-year delay degrees are positively associated with funding for labor services and the NIH, with the former being the stronger contributor by an order of magnitude. See Tables A.46–A.48.

- Visual and Performing Arts degrees, which on a four-year delay are positively associated with funding for elementary, secondary, and vocational education. On six- and eight-year delays degrees are positively associated with funding for the Department of Education. See Tables A.49–A.51.

Disciplines for which funding sources do not explain the majority of variation in degree conferral rates include: Architecture (see Tables A.4–A.6); Business (see Tables A.10–A.12); and Education (see Tables A.19–A.21).

It appears, then, that funding in *a priori* relevant areas has the potential to explain a great deal of the variation in degree conferral rates across a range of disciplines. However, it is noteworthy how seldom NASA funding appears as a significant, positive contributor to conferral rates, standing out as the sole positive contributor or as the strongest contributor in very few cases. This does not provide a secure basis for maintaining that NASA spending or space science funding is a significant overall factor in students' educational decisions. Therefore, we have sufficient cause to suspend judgment on an obligation to increase spending on spaceflight in order to inspire students to enroll in STEM degree programs.

It should be acknowledged these data do not support any robust causal conclusions about the impact of funding and degree conferral rates. In all likelihood, comparing degree conferral rates to federal funding is too blunt a strategy, since there are other relevant factors, including effects of the corporatization of higher education, shifts in state and local education and research funding, shifts in demand from the public and private sectors, etc. Other likely factors include variation in socioeconomic status, in aptitudes and skills, in objects of student curiosity, and in societal, peer, and family pressures, etc. Taking stock of all relevant variables are tasks that surpass my sociological ken.

Space and Science Enthusiasm

So much, then, for the idea that spaceflight activities can be justified through their purported role in inspiring students to earn STEM degrees. What about the other half of the educational inspiration rationale, *viz.*, that space exploration helps to satisfy an obligation to increase the scientific literacy of the general public? Such an argument has an august heritage—consider the following passage from Carl Sagan's *Pale Blue Dot*:

> Exploratory spaceflight puts scientific ideas, scientific thinking, and scientific vocabulary in the public eye. It elevates the general level of intellectual inquiry. The idea that we've now understood something never grasped by anyone who

ever lived before—that exhilaration, especially intense for the scientists involved, but perceptible to nearly everyone—propagates through the society, bounces off walls, and comes back at us. It encourages us to address problems in other fields that have also never before been solved. It increases the general sense of optimism in the society. It gives currency to critical thinking of the sort urgently needed if we are to solve hitherto intractable social issues. It helps stimulate a new generation of scientists. The more science in the media—especially if methods are described, as well as conclusions and implications—the healthier, I believe, the society is. People everywhere hunger to understand. (1994, 281)

Unfortunately, there are almost no data that bear directly on space exploration's influence on the scientific literacy of the general public. So as with spaceflight's impact on STEM degree conferral rates, we ought at least to suspend judgment about the effectiveness of spaceflight's potential to increase the scientific literacy of the general public, and grant that there is no clear basis for an obligation to use spaceflight for such a purpose. Nevertheless, I believe a stronger admonition is warranted in this case. In particular, I will provide a *reductio* of the claim that spaceflight positively influences the scientific literacy of the general public.

Suppose there was a positive causal relationship between spaceflight activities and scientific literacy. We would expect, then, that the level of scientific literacy would vary over time with the scope of spaceflight activities. What can be shown is that no such covariation has occurred, and thus that there is no clear causal relationship between spaceflight activities and scientific literacy. My *reductio* takes the following form:

1. Scientific literacy is correlated with support for space exploration.
2. Support for space exploration has not varied significantly over time.
3. Spaceflight activities have varied significantly over time.
4. Thus, spaceflight activities do not correlate with support for space exploration.
5. Thus, spaceflight activities do not correlate with scientific literacy.

    ----

6. Therefore, spaceflight activities are not causally related to scientific literacy.

All that remains is to establish the truth of the premises of this argument. First, consider the evidence for (1)—that scientific literacy is correlated with support for space exploration. As much was indicated by a small-scale study of undergraduates at Syracuse University:

[W]hile college undergraduates claim to know little about U.S. space exploration, they tend to have positive attitudes regarding NASA. Their scientific literacy levels and public support for space exploration are related and this is most evident in political science and health science majors. It may be that the better educated one is about space science, the more likely he or she is to become an informed citizen who participates in public discourse and is therefore more optimistic and supportive of space science. (Cook, Druger, and Ploutz-Snyder 2011, 51)

According to François Nadeau's assessment of data from the General Social Survey (GSS), Cook, Druger, and Ploutz-Synder's sample is likely representative of the U.S. public with respect to attitudes toward space exploration:

Americans are more likely to favor spending on space exploration if they are more knowledgeable about, and have a greater appreciation for, organized science. This confirms that recent findings by Cook and associates apply not only to a small sample of undergraduate students at Syracuse University, but to the American public at large. When forming their spending preferences on space exploration, scientific literacy seems to help many Americans compensate for a lack of elite cues in society. In addition, Americans appear to favor more spending when they complete more college-level science courses. (Nadeau 2013, 164)

Therefore, there is evidence that scientific literacy correlates positively with support for spaceflight.

Second, consider the evidence for (2)—that support for space exploration has not varied significantly over time. Support for NASA has always been relatively high for a federal agency. According to a 2015 Pew Research Survey, 68% of Americans had a favorable opinion of NASA—second only to the 70% favorability rating for the Centers for Disease Control and Prevention.[9] Meanwhile, according to a series of Gallup Polls from 1990 to 2007 (see Fig. 1.1), an average of 57.6% Americans say that NASA is doing an excellent or good job; 28.5% say NASA is doing only a fair job; and only 7.8% say NASA is doing a poor job.[10] Over the same period scientific literacy has also remained relatively stable (Snow and Dibner 2016, 53–54). Finally, in support of (3), consider that during the same period NASA's budget decreased from about 1.0% of the federal budget to about 0.5% of the federal budget.

Since the scope of spaceflight activities (as measured by NASA's funding) does not correlate with the public's approval of NASA, whereas one's level of scientific literacy does correlate with one's approval of NASA, it would be

9. "NASA popularity still sky-high," http://www.pewresearch.org/fact-tank/2015/02/03/nasa-popularity-still-sky-high/ (accessed 26 February 2018).
10. "Americans Continue to Rate NASA Positively," http://news.gallup.com/poll/102466/americans-continue-rate-nasa-positively.aspx (accessed 26 February 2018).

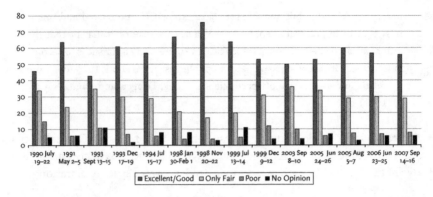

Figure 1.1 Answers by percentage to the question "How would you rate the job being done by NASA—the U.S. space agency? Would you say it is doing an excellent, good, only fair, or poor job?" from a series of Gallup polls from 1990 to 2007.
Credit Line: James S. J. Schwartz.

difficult to maintain that the scope of spaceflight activities has a salient impact on the level of scientific literacy among the general public. Spaceflight activities thus are not effective implements for inspiring the general public to become more scientifically literate, and consequently, we have no obligation to use spaceflight to promote increases in the scientific literacy of the general public.

Things only get muddier when examining support for space science relative to the more general issue of support for science. The common refrain among space advocates that space topics are inherently exciting is not supported by public opinion regarding science funding. A comparison of the GSS natsci,[11] natspac,[12] and natspacy[13] data sets from between 2002 and 2016 reveals that roughly 38% of Americans felt too little was being spent on scientific research, while only 17% felt too little was being spent on space exploration. Meanwhile, about 12% felt that too much was being spent on scientific research, while 33% felt too much was being spent on space exploration. This suggests that Americans are, on average, less sanguine about space exploration than they are about scientific research in general. Relevant here is a recent study which found that respondents tend to overestimate science's share of the federal budget, and that after presented with correct information, there is a significant increase in the percentage of individuals responding that science funding should be increased (Goldfarb and Kriner 2017). This corroborates a

11. GSS Data Explorer, "Supporting Scientific Research," https://gssdataexplorer. norc.org/variables/197/vshow (accessed 17 May 2018).
12. GSS Data Explorer, "Space Exploration Program," https://gssdataexplorer.norc. org/variables/181/vshow (accessed 17 May 2018).
13. GSS Data Explorer, "Space Exploration—Version Y," https://gssdataexplorer. norc.org/variables/199/vshow (accessed 17 May 2018).

suspicion of Alan Steinberg's (2013) that opinions about spaceflight spending are influenced by misperceptions of, and can be improved by accurate communication about, the actual level of spaceflight spending.[14]

Although there is little reason to think that spaceflight activities are effective determinants of scientific literacy, perhaps we could attain increased support for spaceflight (and for science more generally) by increasing overall scientific literacy. The problem here is that, while scientific literacy and support for science are positively correlated, there are many other relevant factors:

> [T]he path from scientific knowledge to positive attitudes toward science or support for science is not always clear. Knowledge affects different subgroups in a population differently depending on a host of factors, including levels of religiosity, political predispositions and worldviews, and deference to scientific authority. These patterns seem to vary depending on the specific scientific issue being explored and the culture in which the data is collected. (Snow and Dibner 2016, 93)

Thus, an increase in scientific literacy, even targeted to a specific domain of science, would not provide a strong guarantee of an increase in support for that domain of science.

LIFE'S BIG QUESTIONS

Another popular rationale for space exploration is that it promises to produce answers to "life's big questions," about, e.g., the origin of life on Earth, or whether terrestrial life is alone in the Universe. Such questions fall under the purview of astrobiology and the scientific search for extraterrestrial life. According to astrobiology advocates, not only are such questions intrinsically worth answering, virtually every person yearns to learn the answers to them. Bruce Jakosky claims as much:

> Astrobiology and astrophysics address questions that are close to universal, both to scientists and to the public. We look for answers to questions about how the universe formed and evolved; how galaxies and stars form, evolve, and die; how planets form, behave, and evolve, and whether they are widespread; whether Earth-like planets exist elsewhere; how life originates and whether

---

14. As Snow and Dibner (2016, 68–70) note, some of the more significant determinants of scientific literacy include socioeconomic status, access to quality education, and other environmental factors. Thus, even if spaceflight had a statistically significant impact on scientific literacy, it would only be a minor factor in a much more complex affair.

microbial life exists elsewhere; and whether intelligent life is unique, rare, or common in the universe. These questions touch us deeply as humans. They get at the basic issue of how we, both collectively and as individuals, relate to our surroundings. (2006, 99)

Two distinct rationales can be distinguished. On the one hand, there is the argument that we have an obligation to answer "life's big questions" via astrobiology and the scientific search for extraterrestrial life because these questions are intrinsically worth answering. On the other hand, there is the argument that we have an obligation to answer "life's big questions" via astrobiology and the scientific search for life because the majority of the public wishes to learn the answers to these questions.

I do not wish to promote skepticism about the first rationale. This is because I would be inclined to agree that questions about the origin and extent of life are intrinsically worth answering—though in Chapter 4 I will question whether the scientific search for extraterrestrial life is all that is important when it comes to environmental protection in space. Nevertheless, we should be wary of the second rationale—that we can found an obligation to answer "life's big questions" on the purported popular support for astrobiology and the scientific search for life. However, we should not dismiss this argument as a mere *ad populum*, because that would ignore the relevance to scientific research of political and societal context. Although we need not go so far as to endorse an extremely democratic position *à la* Philip Kitcher's well-ordered science,[15] nevertheless it is reasonable to assume that in a democratic society at least some research goals should be responsive to the needs and desires of the public. If the public is acutely interested in a particular issue, then that is at least a *prima facie* reason to support research into that issue. It is even more evident that we ought to support research into issues when there is a convergence of public and legitimate scientific interest. Assuming the public is especially interested in the search for extraterrestrial life, and, given that the search is already legitimately scientifically interesting, it seems not at all contentious to claim that there is a *prima facie* obligation to support the astrobiological search for evidence of extraterrestrial life. So, there is some value in estimating the level of the public's interest in astrobiology and the scientific search for extraterrestrial life. And it is on this point that I will raise an objection to the second rationale for astrobiology.

Very little is known about the public's views on astrobiology and the search for extraterrestrial life. I am aware of only one major survey of the American public's attitudes toward extraterrestrial life, and this survey did not attempt to gauge interest in, or support for, astrobiology or the *search*

---

15. See Chapter 3 for more on Kitcher's well-ordered science.

for extraterrestrial life. Although I will discuss the results of this survey in a short while, it should be instructive to form two provisional hypotheses based on well-known aspects of American public opinion which are likely to influence views on astrobiology, *viz.*, the public's views on evolution and on space exploration more generally. As discussed previously, the U.S. public largely approves of space exploration, but perhaps ironically, the U.S. public is generally opposed to funding increases for NASA. It is possible that the U.S. public would react similarly to astrobiology and the scientific search for extraterrestrial life, in so far as they implicate spaceflight. Meanwhile, the U.S. public is very divided with respect to beliefs about the origin of terrestrial life—religiosity is correlated with skepticism about evolution via natural selection. Those skeptical that terrestrial life evolved via natural selection might be similarly skeptical of life evolving anywhere else in the Universe.

Views on Evolution and Religion

Consider first the U.S. public's views on evolution. Among the 35,701 U.S. adults surveyed in Pew Research Center's 2014 Religious Landscape Study, 33% believe that humans evolved due to natural processes, 25% believe that humans evolved "due to God's design," and 34% believe that humans always existed in their present form.[16] This latter figure is down from the 42% who denied evolution in 2005; and the combined figure for believers in evolution (58%) is up from 48% over the same time period. Those believing that humans evolved due to natural processes increased from 26% to 33%; those believing in intelligent design increased from 18% to 25%.[17]

As Table B.1 in Appendix B shows, views on evolution vary considerably by religious tradition. Among those who believe in intelligent design, 90% say their religion is either somewhat or very important in their lives. Of those who believe that humans always existed in their present form, 93% say religion is either somewhat or very important in their lives. Meanwhile, 52% of those believing in evolution due to natural processes say that religion is either somewhat or very important in their lives. Though not asked in the 2014 study, the 2005 study noted that, among those believing that humans always existed in their present form, 63% were "very certain about how life developed" (69% of biblical literalists); meanwhile, 39% of believers in intelligent design were very certain, and only 28% of believers in evolution due to

16. "Religious Landscape Study," http://www.pewforum.org/religious-landscape-study/ (accessed 25 February 2018).

17. "Religion a Strength and Weakness for Both Parties," http://www.people-press.org/2005/08/30/religion-a-strength-and-weakness-for-both-parties/ (accessed 20 February 2018).

natural processes were very certain. Another noteworthy result from the 2014 study is that the more frequently one participates in religious activities (e.g., attending services; prayer; reading or studying scripture), the more likely one is to believe that humans always existed in their present form, or that humans evolved due to God's design. A similar pattern holds for individuals based on the strength of their belief that scripture is the literal word of God.

Now, what does this have to do with support for astrobiology? It must be admitted up front that one's views on evolution may not be predictive of one's views on astrobiology or on extraterrestrial life. After all, people are known to possess inconsistent belief sets. Nevertheless, it is plausible that that skepticism about evolution correlates with a lack of interest in *scientific* answers about, e.g., the origin of life, or the existence of extraterrestrial life. That is, those with extreme doubts about evolution via natural processes might not only lack interest in astrobiology; they might also prefer that astrobiologists not engage in research into the origin and extent of life. This could be evidenced by the fact that the majority of creationists are "very certain" in their creationism, and so presumably are not at all interested in what biologists have to say on the issue. The cogency of this hypothesis depends on the degree to which evolution, astrobiology, and the search for extraterrestrial life are linked together in the eyes of the public—an interesting avenue for further research, to be sure.

Views on Space Exploration

Consider next the public's views on space exploration. As NASA historian Roger Launius (2003) has noted, there has been and there continues to be a mismatch between, on the one hand, the public's approval of NASA and the space program, and, on the other hand, the public's willingness to support funding increases for space exploration. As mentioned earlier in this chapter, Gallup polls from 1990 to 2007 revealed that an average of 57.6% of Americans thought that NASA was doing an excellent or good job; 28.5% thought NASA was doing only a fair job; 7.8% thought NASA was doing a poor job; and 6% had no opinion.[18] According to the General Social Surveys (GSS) from 2008 to 2014, (see Fig. B.1 in Appendix B), an average of 67.2% of Americans were either very or moderately interested in space exploration, while an average of 32.1% were not at all interested.[19] Meanwhile, a different picture emerges when the public is asked specifically about *funding* for NASA. Table 1.1 captures GSS responses from 1973 to 2016 to the question "are we

18. See note 10.
19. "GSS Data Explorer: Interested in Space Exploration," https://gssdataexplorer. norc.org/variables/3459/vshow (accessed 26 February 2018).

Table 1.1 TOTAL RESPONSES TO THE QUESTION "ARE WE SPENDING TOO MUCH, TOO LITTLE, OR ABOUT THE RIGHT AMOUNT" ON SPACE EXPLORATION? PERCENTAGES ARE BASED ON COMBINING TOTALS FROM BOTH THE NATSPAC (1973 ONWARDS) AND NATSPACY (1984 ONWARDS) GSS DATA SETS.

	Since 1973	Since 2000	2016
Too little	13%	16.1%	20.5%
About right	39.5%	42.3%	45%
Too much	40.8%	34.4%	25.4%
Don't know	6.2%	7%	9%

spending too much, too little, or about the right amount on" space exploration? An average of only 13% said that too little was being spent; 39.5% said spending was about right; and 40.8% said that too much was being spent. (Of course, the averages for 2000 onward are friendlier to space funding, and the results for 2016 are even friendlier yet.)[20]

Focusing on the period between 2006 and 2010, William Bainbridge finds evidence of the influence of scientific literacy and religion on beliefs about space funding. Among the 52.7% of GSS respondents who correctly identified that "human beings, as we know them today, developed from earlier species of animals," 18.2% said there is too little space funding; meanwhile, only 10.7% of those denying this claim thought there is too little funding for space (Bainbridge 2015, 122). Similarly, among the 50.4% who said that "the universe began with a huge explosion," 20.2% said there is too little space funding; of those denying this claim, only 9% said there is too little space funding (122). Scientific literacy may be the primary culprit here, since other questions seemingly unrelated to any potential conflict between science and religion revealed similar differences in attitudes toward space. For instance, of

20. See Figs. B.2 and B.3 in Appendix B for a representation of how opinions on space funding have fluctuated over time, both during the entire period of the GSS, and since 2000. These figures also provide a comparison with NASA's share of the federal budget over the same period. Table 1.1 as well as Figs. B.2 and B.3 use data aggregated from both the natspac and natspacy GSS data sets. The sole difference between these questionnaires was that the natspac questionnaire (in use since 1973) asked about funding for "the space exploration program" whereas the natspacy questionnaire (in use since 1984) asked about funding for "space exploration." As Table B.3 in Appendix B shows, from 1984 onwards there is very little difference in the responses between the natspac and natspacy wordings. This means it is likely that the responses during the period from 1973 to 1983 regarding "the space exploration program" are representative of how individuals would have responded during this time had they been asked about "space exploration." Thus, aggregation of the natspac and natspacy responses is appropriate.

those correctly denying that lasers work by focusing sound waves, 20.1% said there is too little space funding; meanwhile, of those incorrectly agreeing that lasers work by focusing sound waves, only 9.8% said there is too little funding for space (122). Nevertheless, Bainbridge does find data that bear directly on the influence of religion, and in particular, on the strength of one's religious convictions. Of those who "know God really exists" and who have no doubts about it, only 11.7% said there is too little funding (124–125). Compare this with 22.4% of atheists and 24.9% of agnostics (125). A similar trend emerges when examining frequency of attendance of religious services. Of those who say they attend religious services more than once per week, only 7% say there is too little space funding; compare this with 16.9% of those who never attend religious services (125).

According to Joshua Ambrosius' analysis of data from the GSS and from several other surveys, we must also be mindful of the influence of religious tradition. Just as religious tradition appears to influence one's views on evolution, so too does it seem "to affect space knowledge, policy support, and the general benefits of space exploration" (Ambrosius 2015, 22). As with evolution, Evangelical Christians stand out:

> Evangelicals are indeed less knowledgeable (even if unwilling to admit their knowledge), interested, and supportive of space/space policy than the population as a whole and/or other religious traditions. This is a problem for the future of space exploration because Evangelicals make up more than one-quarter of the U.S. population . . . and thus a significant share of potential space-minded constituents. (25)

Meanwhile, those identifying as Jewish, Hindu, or Buddhist had greater than average interests in space (23). Thus, Evangelicals may have an outsized impact on the correlation between religiosity and pessimism about space, since Evangelicals are more likely than are other religious groups to attend services once a week or more, to believe that scripture is the literal word of God, etc.

It is plausible, then, that the public largely approves of astrobiology and the scientific search for life, but that like space exploration more generally this support is moderated by scientific literacy, religiosity, and religious tradition, and it does not extend to willingness to provide increased funding for astrobiology projects.

Views on Astrobiology and the Search for Extraterrestrial Life

Given that educational level and scientific literacy positively correlate both with belief in evolution by natural selection and with willingness to increase space funding, and that religiosity negatively correlates with belief in evolution by natural selection and with willingness to increase space funding, it could be that astrobiology and the search for extraterrestrial life are subject to acute levels of disapprobation, at least among those with less education or those with higher religiosity. So, it is not safe to assume that astrobiology and the search for extraterrestrial life inherit the same degree of popularity as space exploration more generally. What of the data that bear directly on astrobiology? To the best of my knowledge there have only been four surveys that have attempted to measure the public's interest in, and willingness to support, astrobiology and the search for life;[21] and only one of these surveyed the American public—Pettinico (2011). This telephone survey, which took place in 2005, used a random sample of 1,000 U.S. adults, making it the largest survey of its kind. It is also noteworthy in being the only survey that attempts to discriminate between belief in extraterrestrial life of various types (e.g., microbial extraterrestrial life, versus plant- or animal-like extraterrestrial life, versus intelligent extraterrestrial life).[22]

In response to the question "do you believe that there is life on other planets in the universe besides Earth?," 60% of the sample said yes, 32% said no, and 8% were not sure (Pettinico 2011, 104). Belief in extraterrestrial life correlated negatively with frequency of religious service attendance: Only 45% of those attending services weekly believed in life on other planets, whereas 70% of those rarely or never attending services believed there was life on other planets (104). Pettinico also reports a positive correlation with belief in life on other planets and household income (106). Of the 32% not open to extraterrestrial life, 56% cited religion as a major reason (113). Among this group (about 18% of the total sample), frequency of attending religious services correlated positively with the identification of religion as a major reason for rejecting the possibility of extraterrestrial life, with 72% of those attending services weekly giving this reason compared to only 31% of those attending rarely or never (114).

This information might lend credence to the analogy between evolution and astrobiology, since religiosity is negatively correlated with belief in evolution by natural selection and with belief in the possibility of extraterrestrial life. Nevertheless, these results do not provide definitive insight into the

---

21. Swami, et al. (2009), Pettinico (2011), Oreiro and Solbes (2017), and Persson, Capova, and Li (2019).

22. More frequent are small-scale surveys testing educational outcomes of including astrobiology in school curricula, e.g., Foster and Drew (2009) and Hansson and Redfors (2013), or surveys estimating the response of religious individuals to the discovery of extraterrestrial life, e.g., Peters (2013). Consequently, the reader should not be left with the impression that there have only been four surveys on astrobiology—just that there have only been four surveys that aimed specifically to identify the public's interest in, and support for, astrobiology and the scientific search for extraterrestrial life. Data from Bainbridge (2011) have been excluded from consideration because they deal primarily with views on extraterrestrial *intelligence*.

public's interest in and support for astrobiology, if only because of a curious spread of beliefs about the likely nature of extraterrestrial life. Of the 68% open to the possibility of extraterrestrial life, 45% think that there very likely is extraterrestrial microbial life; 25% think that there very likely is extraterrestrial life similar to plants; and 21% think that there very likely is extraterrestrial life similar to animals (111). Meanwhile, 30% think that there very likely is extraterrestrial life comparable to humans; and 39% percent think that there very likely are *superior* extraterrestrial intelligences (111). Pettinico ventures an explanation as to why these views do not accord with scientifically informed expectations (that the likelihood of extraterrestrial life decreases as the complexity of that life increases):

> It makes sense that the public feels very simple life forms such as microbes are the most probable extraterrestrial life forms, since most space scientists would generally agree—at least that extraterrestrial microbes would probably be more common than more advanced life forms. However, the public is more likely to believe in the probability of advanced life forms . . . than they are to believe in the probability of plant-like or animal-like life forms. This may be, in part, due to the impact of the media, which tends to emphasize human-like or advanced extraterrestrial life forms. When average Americans think about aliens, they may more easily envision *Star Trek*'s Klingons than they do some sort of lower-level animal. (111)

Of course, it is important to ask why it is that, e.g., media portrayals of extraterrestrial life tend to be extraterrestrial intelligences (ETI), and very often human-like ETI. Clearly, human-like ETI are easier to imagine and to portray in film. But it also could be that ETI, and human-like ETI, are simply more interesting to most people than other forms of extraterrestrial life. Thus, it could be that what drives the responses in the cases of human-like and superior ETI are not scientifically grounded opinions but instead *preferences* based on what the respondents hope is the case or what they would find most exciting. For this reason, it is important when surveying the public to attempt to control for this potential difference in enthusiasm about extraterrestrial life. It is possible that those who are enthusiastic about the search for life are primarily enthusiastic about the potential discovery of human-like or superior ETI, and less so concerning "simpler" forms of extraterrestrial life.

It must be admitted, then, that little is known with any confidence about the public's views specifically about the scientific search for extraterrestrial life as it is currently being conducted, e.g., via robotic exploration of Mars or via exoplanet biosignature detection. There are, in my estimation, five issues that must be addressed in future research before we can draw robust conclusions about the public's views on astrobiology and the scientific search for extraterrestrial life. The first issue is that interests in extraterrestrial life are diverse, and could come from interest in the possibility of microbial life in the Solar System, from interest in the possibility of intelligent life elsewhere in the Universe, or from interest in the paranormal (e.g., UFOs and alien visitation). These interests are independent. A person could be interested in the paranormal and not at all interested in, e.g., microbial extraterrestrial life. Similarly, a person could be very interested in the possibility of life on Mars but not at all interested in possible biosignatures from exoplanets. A second issue is that beliefs about extraterrestrial life are not identical to beliefs about the importance or value of searching for extraterrestrial life. A person's degree of belief in the existence of extraterrestrial life is not tantamount to the degree to which they think it is important to search for evidence of extraterrestrial life; and the latter was not addressed in Pettinico (2011).[23]

A third issue is that interest in extraterrestrial life is not tantamount to interest in what *science* has to say about extraterrestrial life. Some individuals may be interested in the origin of human life, but not at all interested in what evolutionary biologists have to say on the topic. It is plausible that the same holds for extraterrestrial life—that many individuals interested in extraterrestrial life may have little interest in what astrobiology uncovers. Two obvious examples would be conspiracist-types who believe in alien visitation despite the privation of good evidence for such beliefs; as well certain religious individuals who believe (and have little doubt) either that God only created life on Earth, or that God created life wherever it exists. The fourth issue is that interest in extraterrestrial life, and even interest in the science surrounding extraterrestrial life, is not tantamount to willingness to increase funding for the scientific search for extraterrestrial life. If the analogy with space exploration is apt, then we should expect few individuals to be supportive of funding increases for the search for extraterrestrial life, even if most approve of the search. It is important to ask both sorts of questions when soliciting the public's views on the search for extraterrestrial life.

A final issue is that absolute interest in extraterrestrial life is not tantamount to relative interest in extraterrestrial life or to viewing the search for extraterrestrial life as a priority. It is possible that those who are very interested in extraterrestrial life, and even those who think it deserves funding increases, nevertheless do not view the search for extraterrestrial life as a priority compared to their other interests. As much holds for views on space

23. Though respondents' interest in, and willingness to support the search for extraterrestrial life were addressed in Swami, et al. (2009), Oreiro and Solbes (2017), and Persson, Capova, and Li (2019), none of these surveys attempted to control for views about extraterrestrial life versus ETI. It should also be noted that they sampled non-U.S. populations (Austria and Britain; Spain; and Sweden, respectively), limiting their relevance to a comparison with U.S. views on evolution and exploration. See Schwartz (2020) for further discussion.

exploration more generally. Thus, it is not enough to merely ask in isolation whether one thinks the search for extraterrestrial life is important. Rather, the goal should be to determine how important the search for extraterrestrial life is compared both to other space exploration objectives and to other projects, scientific or otherwise.

Admittedly, there is more to astrobiology than the search for extraterrestrial life, and thus, interest in astrobiology may come apart from interest in the search for extraterrestrial life. Nevertheless, the search for extraterrestrial life is a core item on the astrobiology agenda, and one that astrobiologists are not shy about promoting. It is significant, then, that there is no clear evidence of broad, unequivocal public interest in and support for astrobiology and the scientific search for extraterrestrial life—which leaves as unsubstantiated the claim that the public's desire to see astrobiology answer "life's big questions" provides sufficient grounds for the existence of an obligation to support astrobiology in this way.

THE URGE TO EXPLORE

Another way to argue for an obligation to explore space is by assuming that only through space exploration can we satisfy certain innate human urges, e.g., innate urges to explore or to migrate.[24] Human space exploration and space settlement would be the primary activities implicated by such an obligation. One proponent of this rationale is Robert Zubrin, founder of the Mars Society:

> The human desire to explore is thus one of our primary adaptations. We have a fundamental need to see what is on the other side of the hill, because our ancestors did, and we are alive because they did. And, therefore, I am firmly convinced that humanity will enter space. We would be less than human if we didn't. (2000, 275)

Other promulgators include Carl Sagan and Ian Crawford. Sagan writes in *Cosmos* that:

> We embarked on our cosmic voyage with a question first framed in the childhood of our species and in each generation asked anew with undiminished wonder: What are the stars? Exploration is in our nature. We began as wanderers,

24. Thanks to Luke Matthews for discussion concerning genetics and human migration.

and we are wanderers still. We have lingered long enough on the shores of the cosmic ocean. We are ready at last to set sail for the stars. (2013, 206)

Crawford, meanwhile, makes an even stronger claim about the potential necessity of space exploration for the well-being of humanity:

[T]here are reasons for believing that as a species *Homo sapiens* is genetically predisposed towards exploration and the colonisation of an open frontier. Access to such a frontier, at least vicariously, may be in some sense psychologically necessary for the long-term wellbeing of human societies. It is important to note that this is not merely a western predisposition, but a *human* one—one that led to the human colonisation of the entire planet following our evolution as a species in a geographically restricted corner of Africa . . . . [R]egardless of how seriously one takes these arguments, it must be true that our horizons will be broader, and our culture richer, if we engage in the exploration of the universe than if we do not. (2005, 260)

Despite its impressive following, it is problematic to claim that exploratory and migratory tendencies characterize humanity. To begin with, such claims are ambiguous, as they could be interpreted in at least one of three ways: Perhaps such claims refer to the idea that humanity has some kind of a fate or "destiny" that lies in space. Perhaps such claims refer to the idea that exploratory and migratory behaviors are essential features of human cultures. And perhaps such claims refer to the idea that exploratory and migratory behaviors are essential features of humans on an individual, biological level. I call these the mystical, cultural, and biological formulations of the claim that exploratory and migratory tendencies characterize humanity.

If there is truth to at least one formulation of the claim that exploratory and migratory tendencies characterize humanity, then it would be possible to use such a claim as a premise in an argument for an obligation to support those spaceflight activities that satisfy urges to explore or to migrate. To begin by tackling the obvious, even if it is uncontroversial that there is truth to at least one formulation of the claim, we would, as Rayna Slobodian (2015) recognizes, risk the naturalistic fallacy if we infer immediately from one of these formulations that it would be good for humans to act on these impulses. In the very least it would have to be argued that acting on these impulses produces more good than not acting on them—either through the satisfaction of desires or through the realization of good consequences. Consequently, it would be possible to reject this rationale by arguing that insufficient good would come of supporting forms of spaceflight that satisfy urges to explore or to migrate. I shall in Chapters 2 and 3 argue that sufficient good *does* come from the *scientific exploration* of space, so I do not wish to pursue this way of rejecting an obligation to satisfy our alleged urge to explore. Instead I shall argue against the initial premise, that is, I shall argue that there is little relevant truth to any formulation of the claim that exploratory and migratory tendencies characterize humanity. However, considerations of space militate against a thorough discussion of all three formulations. I therefore will relay only my concerns about the third, biological formulation—but see Schwartz (2017a) for detailed discussions of the mystical and cultural formulations. So, what I would like to focus on for the time being is the claim that exploratory and migratory behaviors are essential human traits in an individual biological or genetic sense. For insights here we must turn briefly to psychology, anthropology, and genetics.

Psychology and Curiosity

Curiosity is often cited as a driver of exploration and migration, and it is worth describing briefly a psychological conception of curiosity, since related states and behaviors are implicated in genetic and anthropological discussions of exploration and migration. Although psychology proffers many theories of human curiosity,[25] nevertheless one important finding is that it is highly idiosyncratic. While it may be true in a very general sense that all humans are curious, this curiosity takes many different objects. One distinction worth drawing is that between *cognitive* or *epistemic curiosity* and *sensory* or *perceptual curiosity*. Cognitive curiosity refers to "the desire for new information, while sensory curiosity is the desire for new sensations and thrills" (Reio 2012, 894). A second useful distinction is that between *specific curiosity* and *diversive curiosity*. Specific curiosity refers to "the desire for a particular piece of information, whereas diversive curiosity refer[s] to a general desire for perceptual or cognitive stimulation" (Kidd and Hayden 2015, 450). Thus, learning of a particular person that they are curious conveys rather little information about this person, since they could have specific cognitive curiosity, diversive cognitive curiosity, specific sensory curiosity, or diversive sensory curiosity. Moreover, learning that a person is curious in either of these ways tells us nothing about which items of information or which sensations would help to satisfy their curiosity. The specifics vary considerably from person to person, and there is no evidence that information and sensations related to any one topic or arena—space exploration included— function as common or universal objects of curiosity.

Nevertheless, there is an important connection between curiosity and exploration—but exploration in a strictly psychological and biological sense:

> Exploration entails seeking new information to solve a problem through observation, consultation, and directed thinking (specific exploration) and new

25. See Loewenstein (1994) for a useful review.

sensory experiences and thrills to extend one's knowledge into the unknown (diverse exploration). In a definition that links the two constructs, curiosity is the desire for new information and sensory experiences that motivates exploration of the environment. (Reio 2012, 894)

Thus, curiosity motivates exploration, but generally does not motivate anything so lofty as sending humans to explore the Moon, but usually much more banal acts of information or sensation seeking, such as: tinkering with a new toy; surveying one's local environment (be it one's neighborhood, office, or refrigerator); or experimenting with hallucinogens. To infer from the facts that we are all curious *in some sense* and that we are all explorers *in some sense* to the specific claims that humans generally are curious about and wish to explore *space in particular* would simply be an equivocation. No doubt particular *individuals* are curious about and wish to explore space, but this by no means characterizes the species; to argue otherwise risks the fallacy of composition.

Genetics and Exploratory Behavior

So much, then, for any help from psychology for claims about space as a universal or characteristic object of human curiosity. What about anthropology and biology? Is it not the case that our genetic heritage has been shaped by various migrations—from the out-of-Africa exodus through the settling of the American West? Shouldn't we expect this long history of migration after migration to have culminated in human organisms that are particularly prone to explore or migrate? Interestingly, research has revealed genes associated with migratory behavior. In particular, various polymorphisms of the dopamine D4 (DRD4) receptor have been claimed to be associated with what is called the *novelty seeking* (NS) phenotype, which refers to

a heritable tendency to respond strongly to novelty and cues for reward or relief from punishment, which leads to exploratory activity in pursuit of rewards as well as avoidance of monotony and punishment. (Roussos, Giakoumaki, and Bitsios 2009, 1655)

Novelty-seeking, then, is another term for the kinds of behaviors motivated by the various forms of curiosity previously described. Individuals exhibiting the NS phenotype might exhibit any number of behaviors, whether related to migratory behavior or to more "local" forms of exploration, e.g., surveying local resources. The connection between DRD4 and the NS phenotype has not been conclusively established. Some studies and meta-analyses, including Laucht, Becker, and Schmidt (2006), Munafò, et al. (2008), and Roussos,

Giakoumaki, and Bitsios (2009), have reported a positive association between certain polymorphisms of the DRD4 gene and novelty seeking. However, other studies and meta-analyses have reported no association, including Schinka, Letsch, and Crawford (2002) and Kluger, Siegfried, and Ebstein (2002). As with many phenotype-genotype relationships, comparatively little is known about the effects of the environment on the determination of the novelty-seeking phenotype. Both sex (Laucht, Becker, and Schmidt 2006) and socio-economic factors, especially during childhood (Lahti, et al. 2006) have been identified as potential moderators of novelty seeking. Similarly, it is likely that other genes affect novelty seeking in significant but as yet unknown ways.

A priori, if, as some of these results suggest, there is a positive association between DRD4 and novelty seeking, then it would not be unreasonable to expect a positive correlation between the proportion of individuals with the relevant DRD4 polymorphisms in a population and that population's distance from East Africa. As Roussos, Giakoumaki, and Bitsios note, traits associated with novelty seeking—such as "efficient problem solving, under-reactivity to unconditioned aversive stimuli and low emotional reactivity in the face of pre-served attentional processing of emotional stimuli" may have been beneficial during periods of migration (Roussos, Giakoumaki, and Bitsios 2009, 1658). As much has been corroborated by other researchers. According to Chen, et al. (1999), there is "a very strong association between the proportion of long alleles of the DRD4 gene in a population and its prehistorical macro-migration histories" (317). (Note that 7R is the most common long allele of DRD4.) What explains this association? Chen, et al. propose two possible explanations. One is what I call the "wanderlust" hypothesis, according to which the traits asso-ciated with DRD4 *impelled* migration. The second is what I call the "selection" hypothesis, according to which the traits associated with DRD4 were *selected for* subsequent to migration. According to Chen, et al. there is "little evidence" for the wanderlust hypothesis:

> The rate of long alleles of DRD4 is about the same for immigrants as for their re-spective comparison groups in the home country. These results suggest that the increased rate of long alleles among migratory groups may have been a result of adaptation to the particular demands of migration. (318)

In other words, Chen, et al.'s findings indicate that the 7R variant (along with other long alleles) of DRD4 was not a driver of but instead was *selected for as a result* of migration, for roughly the same reasons as suggested by Roussos, Giakoumaki, and Bitsios:

> As previous research has shown, long alleles (e.g., 7-repeats) of the DRD4 gene have been linked to novelty-seeking personality, hyperactivity, and risk-taking behaviors . . . . The commonality among these behaviors appears to be the

exploratory aspect of human nature. It can be argued reasonably that exploratory behaviors are adaptive in migratory societies because they allowed for more
successful exploitation of resources in the particular environment migration
entails—usually harsh, frequently changing, and always providing a multitude
of novel stimuli and ongoing challenges to survival. (Chen et al. 1999, 320)

Chen, et al.'s preference for the selection hypothesis over the wanderlust hypothesis has been corroborated by subsequent research.[26]

A significant body of evidence exists in support of the idea that traits associated with novelty seeking DRD4 alleles had adaptive value for individuals
belonging to migratory societies.[27] This does not paint a rosy picture for
attempts to justify an obligation to explore or settle space on allegedly innate
exploratory or migratory tendencies. In so far as it may be true that humans
are explorers or prone to migration, either genetically or psychologically, this
means very little for space exploration and migration in particular. We might
all be curious and engage in exploratory behavior, but we do these things idi-
osyncratically. We are not all curious about the same things; and we do not all
explore for the same reasons and in the same ways. Crucially, it is not a universal feature of human psychology or genetics to yearn to explore or migrate
into unknown destinations in space. Though some among us might have one
of the polymorphisms of the DRD4 gene that are associated with prehistoric
migration, stronger evidence exists for viewing these genes as selected for *subsequent to* migration rather than *prior to* it (because it is likely these genes
were adaptive for migrants[28]). This is not incontrovertible proof that DRD4
or some other gene (or complex of genes) was not importantly responsible for
impelling migration, but there is certainly a lack of good evidence insisting
that the possession of this gene impelled any acts of migration. Consequently,
we cannot point to the possession of certain polymorphisms of DRD4 by particular individuals as evidence that the exploration of and desire to colonize
space is in our genes. While it remains possible that future research will reveal
a strong genetic determinant of behaviors such as curiosity about space or a
desire or urge to explore space, for the moment there does not exist any evidence suggesting a specific genetic basis for these behaviors. Therefore, at the

26. Wang, et al. (2004), Matthews and Butler (2011), and Gören (2016).

27. Novelty-seeking behaviors are not the only candidate explananses for why
NS-alleles of DRD4 were adaptive post-migration. According to Ciani, Edelman, and
Ebstein, "the DRD4 polymorphism seems also associated with very different factors,
such as nutrition, starvation resistance and the body mass index" and that "it is possible that these factors alone might have conferred an advantage of selected alleles,
such as 7R, on nomadic individuals compared with sedentary ones" (2013, 595).

28. And perhaps also *maladaptive* for individuals in societies that do not provide
outlets for novelty seeking, which has been proposed as an explanation for ADHD,
substance abuse, and compulsive gambling in modern sedentary societies; see the
references in Roussos, Giakoumaki, and Bitsios (2009).

moment, we ought to reject any rationale for space exploration that assumes otherwise.

NEW FRONTIERS

Perhaps the most persistent feature of space advocacy rhetoric, at least in the United States, is that the "conquering of the space frontier" is necessary to avoid societal stagnation. This sentiment is outlined neatly in one of the rationales for space exploration given by the National Space Society (NSS):

> To provide the human species with a new "frontier" for exploration and adventure, and to thought and expression, culture and art, and modes of government. The opening of "the New World" to western civilization brought about an unprecedented 500-year period of growth and experimentation in science, technology, literature, music, art, recreation, and government (including the development and gradual acceptance of democracy). The presence of a frontier led to the development of the "open society" founded on the principles of individual rights and freedoms. Many of these rights and freedoms are being placed under increasingly stringent limitations as human population grows and humanity moves towards a "closed society," where eventually everyone eats the same, speaks the same, and dresses the same. "Cultures that do not explore, die."[29]

The NSS's position, which is typical of most who promulgate the frontier metaphor, is at heart an analogical argument that pays uncritical homage to Frederick Turner's (1921) "Frontier Thesis"—that the development of American democracy was linked indispensably to the American frontier. The argument is roughly that, just as the "conquering" of the North American west yielded various societal benefits, so too will the human conquest of space realize various social benefits, including: gains in science and technology, the promotion of cultural diversity, and the invigoration of democratic governance. We consequently have an obligation to settle space in so far as doing this will help realize important social goods.

There are two plausible avenues for criticizing this kind of argument for an obligation to settle space. One strategy is to reject the basis of the analogy, Turner's Frontier Thesis—that the American expansion into the North American west was primarily and/or uncontroversially responsible for the realization of societal goods. It should not escape attention that many persons did not benefit from the westward American expansion but in fact suffered tremendously because of it, viz., the natives already inhabiting the North

29. National Space Society, NSS Statement of Philosophy, available at: http://www.nss.org/about/philosophy.html (accessed 21 March 2018).

American continent. This expansion was beneficial almost exclusively to white European settlers and their descendants. Furthermore, contemporary historians tend to be skeptical about the role of the western frontier in the advancement of science, technology, and democratic governance. Developments in these areas are more plausibly attributed to events in the eastern United States. Since professional historians have made this case persuasively,[30] I shall focus on a second avenue of criticism, which casts doubt on the effectiveness of space settlement as a means for realizing societal goods like those adumbrated by the NSS: gains in science and technology, the promotion of cultural diversity, and the invigoration of democratic governance.

Here I will parley mainly with the claim that space settlement is an effective implement for promoting cultural diversity, since I discuss the other two issues in great detail in subsequent chapters. However, it is worth making a few brief remarks about the role of space settlement in science, technology, and democratic governance. With respect to science and technology development, and as I argue in Chapter 3, basic scientific research is one important driver of new knowledge, and in turn, new applications of knowledge. If the claim here is that space settlements will *themselves* become loci of basic research, then we should be skeptical. In the instantaneously lethal environments of space, settlers will have to devote the majority of their energies toward securing and maintaining the basic necessities of life, which will leave little time for developing specializations in scientific research. At least through the early decades of their existence, it will be especially rare to find space settlements containing groups of researchers with overlapping specializations, which will hinder collaborative research (a norm in many areas of science). More plausibly, space settlements would only promote basic research and technology development through the transmission of new data and engineering problems to terrestrial researchers who have the spare energy to analyze these data and solve these problems. So, while there may be some truth to the claim that space settlements will generate new knowledge and technological innovations, we must be careful not to frame this as an intrinsic feature of space settlement. Second, with respect to governance, space settlements are likely to manifest many determinants of totalitarianism. Given that space settlements face *continuous* and *instantaneous* existential threats (e.g., failure of air production, vandalism of life-support systems, accidental depressurization), settlers will experience much greater pressures to abdicate personal liberties and to accept pervasive surveillance, as well as strict, centralized control of the production and distribution of the basic necessities of life. The result could well be a totalitarian settlement, as opposed to a liberal democratic one—of this, more in Chapter 6.

30. For an overview of this response, see Schwartz (2017b); and for the work of historians see Limerick (1992), McCurdy (2011), and Pyne (2006).

What, then, of space settlements' ability to promote cultural diversity? First, consider the reason why "space frontier" enthusiasts are concerned about cultural diversity:[31]

> In the twenty-first century, without a Martian frontier, there is no question that human cultural diversity will decline severely. Already, in the late twentieth century, advanced communication and transportation technologies have eroded the healthy diversity of human cultures on Earth. As technology allows us to come closer together, so we come to be more alike. Finding a McDonald's in Beijing, country and western music in Tokyo, or a Michael Jordan T-shirt on the back of an Amazon native is no longer a great surprise. (Zubrin 2011, 299)

How would "opening the space frontier" reverse these trends?

> If the Martian frontier is opened, however, this same process of technological advance will also enable us to establish a new, distinct, and dynamic branch of human culture on Mars and eventually more on worlds beyond. The precious diversity of humanity can thus be preserved on a broader field, but only on a broader field. One world will be just too small a domain to allow the preservation and continued generation of the diversity needed not just to keep life interesting, but to assure the survival of the human race. (300)

As a first response, it is far from clear that Zubrin's concerns are properly motivated. The examples he provides of cultural homogenization are superficial at best. The mere appearance of American brands in other countries belies the fact that, e.g., McDonald's restaurants in the United States are very different from McDonald's restaurants in China—owing primarily to the cultural differences between the United States and China. Moreover, cultural boundaries are much more elusive than Zubrin seems to believe. It is difficult to describe in a clear and general way how to individuate cultures, and this in turn complicates claims about the extent and dynamics of cultural diversity. Therefore, we should not share Zubrin's particular concerns about declining cultural diversity.[32]

Second, to whatever extent substantive instances of cultural homogenization actually occur, this homogenization is unlikely to attenuate simply because a small percentage of the human population decides to settle somewhere in space. Space settlements will not subdue whichever forces are responsible for the blurring of terrestrial cultural lines (e.g., increasingly efficient and

---

31. The following passage comes from Zubrin, who here advocates specifically for Mars settlement. Nevertheless, Zubrin's remarks here are a faithful proxy for the kind of argument that the NSS would offer in support of space settlement more generally.
32. Thanks to Dovie Jenkins for discussion on this point.

sophisticated communication and transportation systems). Moreover, given that international cooperation is often necessary for the pursuit of ambitious spaceflight projects like space settlement, extended settlement efforts might themselves become significant contributors to cultural homogeneity. If Zubrin is asking us to believe that settling Mars would serve as a prophylactic against terrestrial monoculture, we should find his request troubling. On the other hand, Zubrin suggests that cultural diversity will reappear on a "broader," solar-systemic field. Given the unique challenges and environments space presents, it is certainly possible that space societies will develop culturally in unique and unpredictable ways, and in particular, in ways distinct from any terrestrial culture. But if, as seems likely, there will be significant ideological and technological exchange between terrestrial and space societies— if news sources, scientific findings, fashions, etc., are shared—then space societies may come to resemble terrestrial societies in many ways. Whether this will come to pass is unsettled. Although this uncertainty cuts both ways, it does mark as premature the expectation that space societies will promote cultural diversity in the way that Zubrin envisions.

It is possible for someone in Zubrin's position to reply that, as space and terrestrial cultures eventually co-mingle and homogenize, it will then be incumbent upon us to pursue further space settlements to promote new cultural diversity and to keep Solar System-wide monoculture at bay. But this reply sounds like a proposal to run away from the problem, rather than a proposal to solve the problem. If we are genuinely concerned about preserving cultural diversity, then we should not adopt a policy of avoiding opportunities for cultural exchange. Instead we should strive to learn how to engage in cultural exchange and interaction without either turning into cultural assimilation. Zubrin's entreaty only provides a temporary, if shifting, escape and omits a key factor that remains relevant independent of whether we ever settle space—that if we are committed to preserving cultural diversity, we have to learn how to preserve it on Earth.

The upshot is that we have reason not to share Zubrin's and the NSS's hopes that conquering the space frontier will help to realize scientific and technological advances, promote cultural diversity, and invigorate democracy.[33] That is not to say that we do not have an obligation to realize goods like these, just that space settlement is not at present an effective tool for these tasks. This undermines the argument for the claim that we have an obligation to settle space for the purpose of securing these societal goods.

33. Again, see Schwartz (2017b) for a more detailed discussion of these three issues.

Another common rationale for space exploration[34] is that only by exploiting space resources can we free ourselves from the deleterious effects of terrestrial resource depletion and thus improve human material well-being. A succinct expression of this position is found again the NSS Statement of Philosophy:

> The majority of humanity lives at an economic level that is far below that of the Western democracies. Outer space holds virtually limitless amounts of energy and raw materials, which can be harvested for use both on Earth and in space. Quality of life can be improved directly by utilization of these resources and also indirectly by moving hazardous and polluting industries and/or their waste products off planet Earth.[35]

It is possible to construct an argument for an obligation to exploit space resources that is in the spirit of this passage. First it must be assumed that we are obligated to improve human well-being, to mitigate the effects of terrestrial resource depletion, pollution, etc. Second, it must be assumed that new sources of energy and raw materials are necessary (or are the best means) for satisfying obligations to improve human well-being, etc. And third, it must be assumed that the exploitation of space resources is a necessary (or the best) means for accessing sufficiently rich sources of energy and raw materials without increasing terrestrial pollution. From these assumptions it follows that we have an obligation to exploit space resources.

Those familiar with environmental ethics might note several blind spots in the reasoning used in this argument. One is the second premise, against which one might argue that a better way to improve human well-being is by reducing energy and resource *consumption*, rather than feeding demand by "offshoring" new production to space. If sustainable resource policies can be implemented successfully, then they may be more effective than a policy of space resource exploitation (though these policies are by no means mutually exclusive). A second blind spot, which speaks against the first premise, is that we may have incorrectly judged the source of our obligation to reduce the deleterious effects of resource depletion and pollution. If we take seriously certain non-anthropocentric theories of ethics, then we might hold that an alternative grounding for such an obligation are duties to other living organisms, species, or ecosystems. One might even think it is plausible that with respect to energy, resources, and pollution, nonhuman interests ought to win out over human interests. A third blind spot, which speaks against the third premise, is that "off-shoring" energy and resource production to space does not actually

---

34. Defended by a former version of myself (Schwartz 2011).
35. See note 29.

reduce pollution but merely changes its location. It is true that such pollution would no longer directly affect terrestrial ecosystems, but it could have deleterious effects on spaceflight activities and on space environments themselves. For instance, the proliferation of orbital debris threatens to render popular Earth orbits unusable for generations—a problem known as the "Kessler syndrome." Meanwhile, pollution on lunar and planetary surfaces could be hazardous to those residing in these locations, to the scientific study of these locations, or (if one takes seriously duties to lifeless environments) to these moons and planets themselves.[36]

Meanwhile, those unfamiliar with either astronomy or planetary science might object against the third premise that space does not contain any useful resources. That impression would be far from the truth. Solar energy is abundant in space; and among the resources of the planets, moons, comets, and asteroids is every element needed to sustain orders of magnitudes more humans than are presently living.[37] Nevertheless this is roughly the right place in the argument to lodge a decisive objection. In particular, it should be doubted that space resources can be exploited in ways that *effectively* help us to satisfy obligations associated with human well-being and environmental protection. I shall get into specifics in Chapter 5, where I discuss not only space resource availability, but also policy and regulatory issues related to space resource exploitation; so a very brief summary will have to suffice here.

Although the Solar System contains abundant resources (including water, aluminum, iron, nickel, and platinum-group metals), nevertheless our ability to access these resources will remain exceptionally limited for some time. At the moment humanity is incapable of engaging in any significant exploitation of space resources, although within the next 50 to 100 years it may become technologically feasible to harvest resources from nearby sources, viz., the Moon and near-Earth asteroids (NEAs), i.e., asteroids with perihelia below 1.3 AU.[38] A significant limiting factor is energy cost (i.e., fuel requirements). At energy costs similar to those associated with traveling to and returning from the lunar surface, only a very small percentage of the NEA population can be accessed. Since early exploitation activities will take advantage of the most easily accessed resources, this means that the most attractive NEAs and the surface resources of the Moon will be the first to be exploited, with increasing

36. These matters are taken up in more detail in Chapters 4 and 5, but see also Cockell (2007) and Williamson (2006) for accessible introductions to issues related to pollution in space.

37. I point the interested reader to Lewis (1997), Lewis (2015), and Matloff, Bangs, and Johnson (2014) for surveys of the vast array of resources that are available (in principle) in the Solar System.

38. The perihelion of an object in solar orbit refers to the lowest point in its orbit; one AU (astronomical unit) is slightly under 150 million km, or approximately the average distance from the Sun to the Earth.

difficulties and costs associated with deeper lunar resources and with NEAs with higher energy requirements for rendezvous and return. What's more, due to the nature of orbital mechanics, missions to NEAs will in many cases require years for planning and implementation. In the absence of distant, futuristic technology, space resources will only be delivered to Earth in a slow, staggered trickle (if at all). This does not foretell a time when Earth is free of problems stemming from resource consumption, since space resources are unlikely to arrive in sufficiently large quantities. Quite the opposite—the relative poverty of near-Earth space ought to remind us of how important it is to pursue sustainability initiatives irrespective of how diligently corporations such as Planetary Resources champion the cause of space resource exploitation. "Relief" via space mining cannot be relied upon to make up for terrestrial resource shortages.

At present, then, space resource exploitation is not an effective means for satisfying an obligation to improve human well-being. Consequently, there is no presently existing obligation to exploit space resources. Moreover, as I shall argue over the courses of Chapters 3, 4, and 5, exploiting space resources risks conflicts with a stronger duty associated with space environments, *viz.*, to use these environments for scientific study.

SPECIES SURVIVAL

The final rationale[39] I will discuss in this chapter is that we have an obligation to settle space in order to extend human life. The necessity of space settlement for satisfying an obligation to extend human life is clear over the very long term. Due to the inevitable expansion of our sun into a red giant, there is a finite bound on the amount of time—measured in billions of years—during which Earth will remain habitable. Meanwhile, over shorter timescales, any number of threats might herald humanity's end: meteorite impacts; solar flares; nuclear holocaust; global pandemics; etc. Thus, the establishment of a permanent, self-sustaining extraterrestrial settlement would improve humanity's chances of surviving through a global terrestrial catastrophe, and consequently it appears we are morally obligated to pursue such settlements. I freely grant that over the long term we have an obligation to pursue space settlement to extend human life. What I reject is that this obligation weighs strongly on us in the present day or in the near future. As this is again a topic I will discuss later and at length (in Chapter 6), a summary of my position will have to suffice for the moment.

39. Also defended by a former version of myself (Schwartz 2011).

It is not clear that there is much *urgency* in the need to settle space. This is because in most cases there are alternative—and likely more effective—strategies for preserving humanity in the face of terrestrial catastrophes. For instance, meteorite collisions can be averted by developing planetary defense technologies; resource depletion can be averted by changing consumption habits; and still other problems (e.g., the expansion of the Sun into a red giant) are billions of years in the future. Space settlements might *improve the chances* of human survival, but it has not been shown that the reduction in the level of risk to the species is worth the considerable expense of space settlement. Perhaps some modest degree of societal support should be given to such initiatives, but they should not be prioritized, and nor should we pursue them at the exclusion of cheaper and more technically feasible "local" solutions.

We must also be mindful of tradeoffs within the space environment itself, as a settlement in space could be neither permanent nor self-sustaining if its members did not exploit nearby resources. Space settlement entails actively disrupting and/or destroying space environments. At the moment nearly every space environment is virtually unexplored and thus, virtually pristine. A scientifically inattentive space settlement regime could destroy countless opportunities for the scientific examination of these pristine environments. As I shall argue over the course of Chapters 3, 4, and 6, the extant value of space environments to science is greater than the extant value of these environments as possible sites of human settlements. Where there is conflict between using space for scientific study and using space for human settlement, we should side with the former (at least for the time being). That is, any duty we have *at present* to use a particular space environment for human settlement is overridden by a duty to engage in the scientific exploration of that space environment.

SPINOFFS AND SCIENCE?

It would appear, then, that neither of the previously discussed rationales for space exploration is particularly compelling. One theme of this assessment is that obligations to support space exploration only attach to *particular* spaceflight activities or objectives, and not to space exploration *simpliciter*. A second theme is that at present, our capabilities in space are quite limited, and we simply cannot effectively satisfy most of the duties attached to particular spaceflight activities (e.g., space settlement, space resource exploitation). A third theme is that not all spaceflight objectives and activities are on a par. As I shall argue, taken together these themes support prioritizing a particular set of spaceflight objectives and activities, *viz.*, objectives and activities associated with the *scientific exploration* the Solar System. But establishing this is not an easy task, for the simple reason that it is not obvious why scientific exploration is valuable in the first place. The task of the next two chapters will be to provide two kinds of reasons why scientific exploration is valuable. One of these reasons, which is a more sophisticated version of the "spinoff" justification for spaceflight, holds that scientific exploration leads to scientific change, and that scientific change leads to material improvements in human society. While I defend this view in Chapter 3, I also hold out hope for the idea that the knowledge and understanding acquired by scientific exploration are valuable in themselves. To make a case that scientific knowledge and understanding are worth acquiring for their own sake we must take a long, but hopefully rewarding detour through epistemology and ethical theory—and it is to this task that I now turn.

CHAPTER 2

The Intrinsic Value of Scientific Knowledge and Understanding

If the reasoning of the previous chapter is broadly correct, then many widely promulgated rationales for space exploration are unimpressive. In the present day, a strong ethical justification for spaceflight cannot be founded on duties associated with promoting scientific literacy, curiosity about extraterrestrial life, urges to explore, etc. This raises the question of whether any spaceflight objective is subject to a compelling moral defense. As I argue through the course of this chapter and the next, we have a collective or societal obligation to acquire scientific knowledge and understanding, and this obligation provides the basis for a duty to engage in the *scientific exploration* of the Solar System. However, it is not immediately obvious why we should be obligated to acquire scientific knowledge or understanding. And even if we have such an obligation, its existence is hardly helpful for determining how we should apportion society's limited intellectual and material resources to scientific research. It is similarly open how we should apportion the resources dedicated to scientific research to the many and varied scientific research programs. If none of these issues can be addressed, then the project of establishing an obligation to support *space* science would appear intractable.

My overall argument comes in four parts. In the first two parts I lay out two kinds of arguments for the existence of an obligation to engage in scientific research: that this research produces knowledge and understanding that are both intrinsically and instrumentally valuable. In the third part I provide criteria for applying this obligation to particular research programs. Finally, in the fourth part I show that much of space science satisfies these criteria. The upshot is that there is an obligation to pursue many forms of space research. The first part of my overall argument—establishing the intrinsic value of scientific knowledge and understanding—is undertaken here, with the second, third, and fourth parts reserved for Chapter 3.

I begin in Skepticism about the Value of Science by responding to Lars Bergström's skepticism about the intrinsic value of scientific knowledge. Bergström's key objection is that since we do not have an *overriding* obligation to acquire scientific knowledge, this knowledge therefore lacks intrinsic value. I argue that his objection rests on mistaken assumptions about the relationship between intrinsic value and moral obligation. This motivates the discussion in Intrinsic Value: General Considerations of my approach to the justification of attributions of intrinsic value, which is broadly naturalistic in character: Something is intrinsically valuable when making such an assumption factors in the best explanation of a legitimate practice under our scientific worldview. For example, that it is intrinsically valuable to acquire understanding of mathematics factors in the best explanation for why so many mathematicians pursue this understanding for its own sake.

In Intrinsic Value and True Belief I consider various arguments for the intrinsic value of true belief, ultimately arguing that the acquisition of a great variety of true beliefs counts as intrinsically valuable according to the approach outlined in Intrinsic Value: General Considerations. The two subsequent sections (Knowledge and Understanding; Understanding and Cognitive Achievement) consider whether two alternative epistemic states, *viz.*, knowledge and understanding, have a distinctive value when compared to true belief. In Knowledge and Understanding I indicate why we should follow certain epistemologists in holding that knowledge has not been shown to be more intrinsically valuable than true belief. Thus, the intrinsic value that arises in cases of knowledge is simply the intrinsic value of true belief. However, as I discuss in Understanding and Cognitive Achievement, there are accounts of the nature of understanding according to which understanding has a distinctive (though *non*-epistemic) intrinsic value. So the pursuit of scientific *understanding* is of greater overall value than the mere accumulation of knowledge or of true beliefs. I close the chapter in Virtue Ethics to the Rescue? by discussing how scientific understanding (and its value) factors in virtue-based approaches to value and human flourishing.

Before beginning in earnest I should provide two notes to the reader. First, this is the most philosophically challenging chapter of this book, for in the sections Intrinsic Value and True Belief, Knowledge and Understanding, and Understanding and Cognitive Achievement I engage with difficult literature from the epistemic value debate in contemporary epistemology. This is a requirement of providing a substantive, philosophically satisfying defense of the claims that scientific knowledge and understanding are intrinsically valuable. While I have made efforts to avoid being over-technical, I acknowledge that these sections might be of limited appeal to those without philosophical training, or to those interested primarily in the instrumental value of science.

Thus, following Intrinsic Value: General Considerations, non-specialists or those uninterested in the epistemic value debate may wish to skip to Virtue Ethics to the Rescue?, which contains a more accessible discussion of the way in which true beliefs and understandings contribute to human flourishing.

Second, a note on axiology: The term "intrinsic value" has a variety of uses in philosophy. Rather than surveying all possible uses I will simply stipulate what I mean by the term. When I say that something is intrinsically valuable, I mean that, *ceteris paribus*, it adds value to the world simply by existing. In practice, that something has intrinsic value is a *prima facie* reason for adopting various "pro-" attitudes toward it (e.g., for preserving or protecting it, or for thinking it is the kind of thing that ought to exist). Meanwhile, when I say that something is instrumentally valuable, I mean that it is valuable as a means to something else. In practice, that something has instrumental value is only a sufficient reason for taking it into consideration in moral deliberation if it is a means to something else already known to be intrinsically valuable. Thus, learning that something is intrinsically valuable provides this entity with a seat at the table of ethical deliberation—its voice must be heard, and its value must be taken into account. However, it is also the case that intrinsic values are never infinite and that they come in degrees—some entities have more intrinsic value than others (as an original Bonestell painting arguably has more intrinsic value than a toddler's sketch of their parents). At the table of moral deliberation some speak with loud, commanding voices, while others speak with only the faintest of whispers. The possession of intrinsic value is not a magical ethical wand that ends or resolves moral deliberation. Intrinsic values can and often do come into conflict, often without the promise of easy, universally satisfactory resolutions. As the late Larry Powers was fond of saying: Philosophy is hard![1]

1. My focus in this chapter will be on the *justification* of attributions of intrinsic value as opposed to providing an *analysis* of the concept. However, associated metaphysical details should be acknowledged: My conception of intrinsic value (at least when treating knowledge and understanding as bearers of intrinsic value) resembles G. E. Moore's (1903) conception of intrinsic value more closely than it resembles Immanuel Kant's (1993) formulation. That is, I use it as a notion that attaches to entities of various kinds, including states of affairs, and as a notion that can undergird duties or obligation *involving* entities which have this kind of value. Thus, one could say that when an artwork is intrinsically valuable, we have certain duties *involving* this entity but never *for its own sake*, such as a duty to preserve it, to appreciate its beauty, etc. These are duties associated with preserving and appreciating the *value* of the artwork, i.e., the good it adds to the world simply by existing. Meanwhile, it would be difficult to maintain that we have duties *to respect the interests* of artworks, or that artworks have *rights* or a moral *dignity*—all of which follow from the possession of intrinsic value in Kant's sense, which is a kind of value Kant takes to inhere only in rational agents capable of moral deliberation. Whether and to what degree each conception is useful in ethical theorizing and deliberation is a matter of dispute—see, e.g., Anderson (1993), Bradley (2006), Scanlon (1998), and Zimmerman (2001). My own view is pluralistic—that we require a Moorean conception of intrinsic value to make

SKEPTICISM ABOUT THE VALUE OF SCIENCE

A necessary condition on there being an obligation to engage in scientific research is that this research is either good in itself, or produces things (e.g., knowledge, understanding) that are good in themselves. However, according to the heterodoxy advanced by Lars Bergström, neither is science an effective producer of knowledge, nor is scientific knowledge intrinsically valuable. Why does he believe that science is not an effective producer of knowledge?

> [S]cientific theories have become increasingly difficult to understand. Most people may be able to grasp the principles behind the steam engine, for example, but only a small minority understand the functioning of a laser or a computer. Quantum theory is certainly more difficult than Newtonian mechanics. Hence, it may be safely assumed that educated people have never before been as ignorant of the science of their time as they are today . . . . [B]ecause of the scientific progress, there is more ignorance than before. If ignorance is bad, this is an unfortunate effect of science. (Bergström 1995, 506–507)

Bergström's argument is, roughly, that since there is nowadays a greater gap between the average educated person's knowledge and the total knowledge accumulated by humans, then the average educated person is now more ignorant (of the science of their time) than ever before. Since ignorance is bad, it follows that this is an undesirable consequence of science.

This criticism rests on an equivocation. We should distinguish between two ways in which someone might be ignorant: being ignorant *simpliciter* and being ignorant *with respect to some object*. A person is ignorant *simpliciter* when they fail to acquire some minimum breadth and depth of accessible knowledge. That is, when they know less than they ought to know in order to successfully navigate life. This sort of ignorance is clearly bad. Meanwhile, someone is ignorant *with respect to* some thing or topic when that person knows little to nothing about that thing or topic. I am for instance ignorant with respect to the sociodemographics of Wichita, Kansas in the 1930s, since the only thing I know about 1930s Wichita is that the city was home to the first electric guitar performance. There is nothing morally repugnant about being ignorant with respect to a vast array of things or topics.

The only sense in which it might be reasonable to claim that science has increased ignorance is *with respect* to many things, as opposed to ignorance

---

sense of values and obligations associated with entities like artworks, knowledge, relations, and states of affairs *because of the good they bring into the world*, and that we also require a Kantian conception of intrinsic value to make sense of values and obligations associated with persons and with other entities that ought to be treated in certain ways for their own sakes.

simpliciter. The growth of specialized knowledge is not tantamount to a decrement of the knowledge held by the average person. But even here it would be misleading to say that there has been a *growth* of ignorance *with respect to* these specialized topics. Though the *contrast* between specialists and laypeople may have become starker, the fact that now there are specialists whereas before there were not indicates that now there are *fewer* individuals who are ignorant *with respect to* these topics. Thus specialization does not increase but actually reduces ignorance *with respect to* things, at least when measured absolutely. In either case, there is nothing morally repugnant about the way in which science has contributed to "ignorance." To be sure, one could argue that the standards for avoiding ignorance *simpliciter* are positively correlated with the extent of total human knowledge, and that persons today must know more to avoid being ignorant than persons 50 or 100 years in the past. But Bergström offers no evidence that, as a result of science, more persons are ignorant in this way than in the past. That is not to say there is no reason to be concerned about the current state of education in the United States or elsewhere—just that any problems here aren't obviously related to the specialization of science in an ethically salient way.

So much for Bergström's concerns about the knowledge production potential of science. What about his objection to the intrinsic value of knowledge? His discussion begins with an assessment of a W. D. Ross-inspired argument for the intrinsic value of scientific knowledge, *viz.*, that between two otherwise identical universes, we would prefer the universe with more scientific knowledge (Bergström 1987, 54). He observes correctly that such an argument would not convince those "who are not already convinced that knowledge has intrinsic value" (55). Thus it is open which choice is correct when it comes to the comparative value of these two universes. Bergström hopes to show not only that those who prefer the universe-*cum*-knowledge are in the wrong, but that those who see no salient difference between the two universes are in the right.

Bergström first dispenses with two kinds of trivial attempts to establish the intrinsic value of knowledge. One of these attempts identifies the value of knowledge with the value of what is known (56). To modify his example somewhat, suppose a person knows that they are happy if and only if they are happy. Then *ceteris paribus*, the universe in which they know that they are happy contains more good than the universe in which they do not know that they are happy. But this difference in good seems attributable primarily, if not exclusively, to the additional happiness contained within this universe, as opposed to the additional knowledge contained within this universe. What is of interest, then, is whether knowledge *itself* sometimes has intrinsic value, and not whether it is sometimes *materially coincident* with value. A second trivial attempt is to identify the value of knowledge with the value of knowing (57). Knowledge, Bergström acknowledges, often brings with it a sense of satisfaction. This would, as he observes, "only imply that knowledge is extrinsically or instrumentally good" (57).[2]

Bergström next attempts to undercut another possible means for establishing the intrinsic value of knowledge—that this knowledge is desired for its own sake. He is skeptical that anyone ever genuinely desires knowledge for its own sake:

> If some person, X, desires the knowledge of some fact, p, for its own sake, what is it that X desires? Is it (a) the state of affairs that someone knows that p? Or (b) the state that as many people as possible know that p? Or (c) the state that X himself knows that p? Or is it perhaps (d) the state that X himself is the discoverer of the fact that p? (1987, 59)

In his discussion, Bergström claims that we hardly ever desire states (a) and (b), and that instead we desire states (c) and (d), which implies that we do not desire knowledge for its own sake "but rather for the sake of some anticipated consequences" (59). Whether he is correct on this point is a matter that cannot be decided without empirical research. (Anecdotally, I sometimes find myself desiring state (a)—that certain facts become known by someone at some time. But Bergström might explain away this desire as an example of the *instrumental* value of knowledge, *qua* vicarious satisfaction of curiosity.) He does not rest his case here, perhaps in recognition of the evidential weakness of this particular criticism.

His *coup de grace* comes in the form of a thought experiment:

> Jones is suffering from a fatal disease which will kill him in a week or two. He is in bed, but he can read, and the question is what he should read. There are two books available, viz. a novel by P. G. Wodehouse and a text-book in molecular biology. Jones wants to read the novel, and he would in fact enjoy this more than the text-book, which he does not want to read . . . . On the other hand, the biology book would give Jones a great deal of knowledge, much more than the novel. Mrs. Jones knows all this, and she can in fact make Mr. Jones read either of the two books . . . . Now, it seems that, according to the doctrine that knowledge is intrinsically good, Mrs. Jones ought morally to tell Mr. Jones to read the biology book. And the poor Mr. Jones ought morally to read the book even though he does not want to do so. But these conclusions are surely absurd. (62–63)

2. See Kvanvig (2013) for an argument that knowledge and understanding owe their value at least in part to their ability to bring satisfactory closure to inquiry.

While I agree that such conclusions are absurd, Bergström fails to locate the source of the absurdity. There are two assumptions working in this thought experiment. One is that knowledge is intrinsically good. A second is that if knowledge is intrinsically good, then "some action [knowledge acquisition] would be morally obligatory, which is clearly not morally obligatory" (62). Bergström accepts the second of these premises, and thus, in the spirit of the *reductio*, he rejects the first premise. But such a move is overhasty, since the absurdity does not actually follow from these two premises. This is because it is possible to override moral obligations, for instance, if in a particular situation there are other duties which weigh more strongly on an individual. The example of Mr. Jones is one such case, where greater good would be realized by Jones enjoying the short remainder of his life by reading the novel.

However, perhaps by "moral obligation" Bergström means "*absolute* moral obligation," and thus when he claims that Mr. Jones has an obligation to acquire knowledge by reading the biology book he means that Mr. Jones has an *overriding* obligation to acquire knowledge in this way. Then it would indeed follow that, in Bergström's example, Mr. Jones has such an obligation. But it is absurd to think that an obligation to acquire knowledge is the kind of obligation that is absolute or overriding. (That Bergström appears to think so might stem from an equally problematic assumption about intrinsic value—that intrinsically valuable entities should be treated as having overriding or infinite value.) An obligation to acquire knowledge, if there be one, is more sensibly categorized as a *prima facie* duty: Knowledge is an intrinsic good that we have an obligation to realize when doing so does not conflict with the satisfaction of other, more pressing duties. Otherwise we ought never to eat or drink, lest we miss an opportunity to acquire knowledge![3]

Bergström's criticisms, both of the productivity of science and of the intrinsic value of knowledge, are therefore unpersuasive. Nevertheless, as a result of the latter of his criticisms we are momentarily bereft of a coherent argument supporting the claim that knowledge is intrinsically valuable, since quite clearly, it is not acceptable to *assume* that knowledge is intrinsically valuable in isolation from any other considerations. What, then, can be said in favor of the claim that knowledge (or understanding) is intrinsically valuable?

3. Meanwhile, for those who reject the moral pluralism implicit in talk of *prima facie* duties, then it would be possible to construct either utilitarian or deontological analyses that recommended courses of action other than Mr. Jones's reading of the biology book.

Though Bergström's arguments are unconvincing, nevertheless they are interesting because they challenge a fundamental presumption, not just of epistemology—the area of philosophy most directly concerned with knowledge—but of philosophy and of science, and indeed of inquiry more generally, which is that items such as true belief, knowledge, and understanding are of great, freestanding importance. Upon reviewing the epistemic value literature, one could hardly be blamed for coming away with the impression that questions about the value of true belief and knowledge hit upon one of the most fundamental of philosophical bedrocks. But if we rest with brute intuition when it comes to knowledge and its value then we have no means to respond to the likes of Bergström who possess contrary intuitions. Clearly, a more circumspect strategy is called for. But there are a variety of challenges to overcome. First, it is not obvious that all true beliefs are *equally* valuable. Setting instrumentality aside, it would seem that some true beliefs are genuinely significant (e.g., true beliefs about the origin of life) whereas other true beliefs are of little to no value whatsoever (e.g., true beliefs about how many blades of grass there are on the campus of Wichita State University). If we reject the claim that *all* true beliefs are *equally* valuable, then we need some way to make comparative value judgments about true beliefs. And, while it seems good to have certain true beliefs, it seems even better to have knowledge, and even better yet to have understanding. Do knowledge and understanding add value to true beliefs? If so, how? And finally, what has any of this to do with *scientific* true beliefs, knowledge, and understanding?

What I argue for is the claim that at least some epistemic states have intrinsic value. An epistemic state is, roughly, a state in which a cognitive agent bears some kind of epistemic relation to an object of belief or understanding. Examples include truly believing that p; having a justified true belief that p; knowing that p (where p is a proposition); and understanding p (where p could be a proposition, theory, event, etc.). I begin in this section by motivating and describing a view according to which at least some true beliefs are intrinsically valuable. In later sections I survey attempts to show that both knowledge and understanding are more valuable than true belief. Between knowledge and understanding it is the latter that stands the best chance of exhibiting greater value than true belief. In particular, I concede to proponents that understanding includes a component—the intrinsically valuable deployment of cognitive skills—that is not required for mere true belief. Nevertheless I shall argue that it is open whether understanding manifests a distinctive *epistemic* value beyond the epistemic value of true belief.[4] This is followed by

4. See the first paragraph of Understanding and Cognitive Achievement for the relevant definition of "epistemic value."

a discussion of the role of virtue epistemology in bolstering claims about the intrinsic value of true belief and of cognitive skill.

Before getting to any of this I should first describe the approach I intend to take for the justification of attributions of the sort of intrinsic value that epistemic states possess. My perspective is broadly naturalistic in the sense that it begins from the observation of the centrality of the scientific enterprise to us, given the kinds of beings we are, and living the kinds of lives that we live. I see no need to defend scientific methodologies or the scientific worldview against criticism from "outside." I simply do not know how to argue with someone who insists that all of my standards of evidence are inadequate—no matter how reliable, well-confirmed, or successful. That is not a battle I can win, but then it is not a battle I think is worth fighting. I take this account as given, and see no need to question it from without—though legitimate criticism is always possible from within.[5]

Thus, my discussion of intrinsic value, including the intrinsic value of scientific knowledge and understanding, *begins* from a perspective that (in terms familiar to readers of W. V. Quine) recognizes that the scientific enterprise is central to the prediction and control of sense data. My subsequent aim is to articulate what should be said about the value of epistemic states such as true belief, knowledge, and understanding *within* the account of the universe provided to us by the scientific enterprise—construing this enterprise as broadly as can be coherently maintained.

Suppose, then, that some coherent practice factors legitimately in our scientific worldview. Suppose further that certain assumptions about what is intrinsically valuable are part of the best explanation of this practice, its methodology, and its role in our scientific worldview. What I hold is that this counts, *under the scientific worldview*, as a sufficient reason for believing that these assumption are true (or are as close to the truth as we can get). And frankly, I am unaware of what higher bar might be met under the scientific worldview, at least concerning claims that are not subject to direct empirical verification (which includes most (if not all) of theoretical physics and pure mathematics!). As stated previously, I am naturalist enough to reject the skeptical demand that we must justify the scientific worldview from the outside—there is no escape from Neurath's ship. To whatever extent that the scientific worldview entitles us to describe our beliefs as "true," we are similarly entitled to judge as true claims about intrinsic values whenever these claims form part of the best explanation of legitimate practices.

5. This perspective draws inspiration from Penelope Maddy's "Second Philosophy." I point the reader to Maddy (2007) for a very thorough elaboration and defense of this version of naturalism. I do not fully endorse Maddy's position because I believe that it is too methodologically restrictive—see Schwartz (2013a, ch. 5).

One example of a legitimate practice is ethical deliberation. An examination of this practice will quickly uncover patterns of reasoning and behavior that can best be explained by the assumption that human persons are intrinsically valuable. Making some such assumption is *essential* for explaining and engaging in sound ethical deliberation.[6] From this vantage point, skepticism about the intrinsic value of human persons is wholly unmotivated, and unless there emerge legitimate reasons *from within* for revising the practice of ethical deliberation, such skepticism is unmotiva*table*.[7]

Does this perspective risk countenancing too much as intrinsically valuable? The requirement that a practice factors legitimately in our scientific worldview prevents anti-scientific practices (e.g., astrology, spiritualism) from hypostasizing intrinsic values. For an entity to be intrinsically valuable, that entity's existence must be a nomological possibility. Precisely what separates science from non-science from anti-science is a thorny debate that I have no delusions about settling here. I acknowledge that the approach I am offering fares no better or worse than does naturalism more generally with respect to demarcating science, non-science, and anti-science. Thus, my intention is to be no more generous about intrinsic value than naturalism is regarding which practices are constitutive of and compatible with the scientific worldview. But perhaps my approach is not generous enough? After all, aesthetic judgment constitutes an example of a practice that probably lies at the fuzzy periphery of practices which factor legitimately in our scientific worldview. I prefer to respect common intuitions about the intrinsic value of, e.g., artworks. So I would happily concede, *modulo* reflective equilibrium, that practices which exist coherently alongside scientific practices such as aesthetic judgment can also make contributions to the class of intrinsically valuable entities. This reply is no doubt unsatisfyingly brief, and, were my goal to defend a full account of intrinsic value, I would be obligated to say much more about the necessary and sufficient conditions of legitimate practices. However, since my goal here is to examine attributions of intrinsic value stemming from practices (*viz.*, space science) that fall clearly *within* the scientific enterprise, I shall not be detained by issues of demarcation.

6. While here my focus is on a Moorean conception of intrinsic value (see note 1), the same strategy could be deployed in defense of the idea that humans are intrinsically valuable in Kant's sense: That rational agents have a dignity seems baked into the practice of ethical deliberation—we cannot make sense of many legitimate and widely endorsed ethical intuitions *without* assuming that rational agents have a dignity and should be respected as ends in themselves.

7. The debate between Kantians and Utilitarians about whether it is rationality or the experience of pleasure that is intrinsically valuable would be one example of a debate from *within* ethics about which intrinsic value claim forms part of the best explanation of the practice of ethical deliberation.

INTRINSIC VALUE AND TRUE BELIEF

Under the account of intrinsic value on offer, are true beliefs intrinsically valuable? If so, is every true belief intrinsically valuable, or do some true beliefs lack intrinsic value? In order to make some headway here it will be instructive to consider the dispute between those who maintain that all true beliefs are intrinsically valuable and those who argue that not all true beliefs are intrinsically valuable. Notable members of the former group include Jonathan Kvanvig (2008) and Michael Lynch (2004); notable members of the latter include Alvin Goldman (1999), Ernest Sosa (2001), and William Alston (2006). I shall begin by motivating skepticism about the idea that true beliefs are intrinsically valuable without exception. I then consider Kvanvig's attempt to overcome this skepticism. Though I ultimately reject his arguments, his discussion does broach the idea that certain true beliefs might have some kind of non-instrumental *significance* or *importance* not shared by others. I believe it can be argued that these significant or important true beliefs are bearers of intrinsic value, and therefore at least some true beliefs are intrinsically valuable under the inference to the best explanation style argument presented. However, I want to wait on an outline of an *ethical obligation* to acquire true beliefs (at least within the scientific domain) until the final section of this chapter, which takes advantage of certain conceptual resources that will not become available until after my discussion of the value of knowledge and understanding in the sections Knowledge and Understanding and Understanding and Cognitive Achievement.

Why, then, might one doubt that all true beliefs are intrinsically valuable? Because there appear to be many "pointless truths," i.e., truths that, in addition to being of no practical value, appear to have no value of any other kind. Consider the large number of facts asserting the distance between every possible pair of hairs on my cat Elara. If every true belief was intrinsically valuable, then we ought to maintain that I would realize something of value if I acquired any true beliefs about the distances between Elara's hairs. But, as Sosa claims, this is absurd:

> At the beach on a lazy summer afternoon, we might scoop up a handful of sand and carefully count the grains. This would give us an otherwise unremarkable truth, something that on the view before us is at least a positive good, other things being equal. This view is hard to take seriously. The number of grains would not interest most of us in the slightest. Absent any such antecedent interest, moreover, it is hard to see any sort of *value* in one's having that truth. (2003, 156)

Sosa's position, then, is that we do not value the truth *as such*, but instead, that we value having true beliefs that answer questions which are *of interest* to us (158). If a given fact is not of interest to us (because we are not curious about it, it does not serve any role in practical deliberation, etc.), then there is simply no value in acquiring a true belief about this fact. It must be admitted that it is not difficult for any truth to become of interest to us, either as an object of curiosity or as a factor in practical deliberation. For instance, according to the gap theory of curiosity, becoming aware of a gap in one's knowledge can give rise to an impulse to close that gap by acquiring relevant (and ideally true) beliefs.[8] I might have no prior interest in knowing the names of the Apollo 11 backup crew. But when attending trivia night at the bar, I become very curious (to the point of looking up the answer on my phone) in between submitting my team's guess and learning the answer from the host. Afterward I return to my prior state of lacking interest in the topic. Moreover, there is no end to the possible situations in which apparently pointless truths become relevant to practical deliberation, e.g., if an evil interlocutor threatens violence against me if I do not report to him the correct distance between two particular hairs on my cat. But in unexceptional situations, we should agree with Sosa that facts about grains of sand and cat hair simply aren't of interest. Does that prove that true beliefs about these facts are of no value?

It is true that a lack of interest in a given topic, even a universal lack of interest, does not disprove the intrinsic value of true beliefs about that topic. But insisting otherwise would be a curious kind of objection to raise in the present setting. Since intrinsic value cannot be measured via direct empirical means, attributions of intrinsic value always have to be judged indirectly. As I have outlined, the best we can hope for regarding the justification of attributions of intrinsic value uses an inference to the best explanation. Consider again the uncontroversial attribution of intrinsic value to human persons. Why do we accept that human persons are intrinsically valuable? Well, for one reason, it very strongly *seems* to us as though this is the case. But a second and more theoretically informed reason is that the claim that human persons are intrinsically valuable features centrally in most cogent ethical theories, despite great variance among the other details of these theories. The best explanation for these observations is that human persons are indeed intrinsically valuable. Because this theorizing is legitimate, then, we ought to accept that human persons are intrinsically valuable. A similar defense is available for the intrinsic value of true belief, as Kvanvig recognizes when explaining why true belief seems to him to be more valuable than empirically adequate belief:

8. See Loewenstein (1994) for a psychological discussion of the gap theory and other accounts of curiosity.

Should a critic press us on this point, we will find ourselves in an awkward position, for when pressed to account for the intrinsic value of anything, it is very
hard to know what to say . . . . I claim that having the truth is preferable to that
which is merely empirically adequate, and if pressed on this point, I can do little
else than resort to possible cases in which one learns that one's beliefs are empirically adequate but untrue and ask whether readers share my reaction to such
cases; which involves a negative affective sense of having been duped . . . . [T]he
conclusion that truth is intrinsically valuable is the best explanation of the data
before us. (2003, 42)

Asking for more would be out of place, since it is not clear what further evidence even could convince someone still skeptical about the intrinsic value of
an entity. One might for this reason be skeptical about intrinsic value more
generally, but that would be a very different kind of objection.[9] Of course,
what Sosa hopes to have shown is that intuitions about the intrinsic value of
all true beliefs can be explained away, and that what we value is not the truth
as such, but true answers to questions of interest. If he is right about this,
then we are not straightaway interested in the truth, which undercuts the idea
that the intrinsic value of true beliefs provides the best explanation for our
apparent concern for the truth.

What kind of reply can be made at this stage? Kvanvig argues that we still
have reasons to regard each true belief, including the pointless ones, as intrinsically valuable. Kvanvig's strategy is to show that pointless truths are
valueless *only in appearance*, and that on closer inspection each true belief
possesses at least some intrinsic value. He distinguishes between two ways in
which a true belief might appear to lack value (Kvanvig 2008, 203–205): On
the one hand is the idea that the value of a true belief could be *overridden*. The
disvalues associated with *acquiring* or *using* some true beliefs might balance
or override the intrinsic value of *having* those beliefs. For instance, having a
true belief about the contents of a lover's private emails might produce some
small amount of intrinsic "true belief" value, but the violation of trust and
privacy would create an overall disvalue. Varying the details of the case could
lead to different situations where either more overall disvalue, less overall disvalue, no overall value, or positive overall value is created. On the other hand
is Kvanvig's idea that the value of a true belief could be *undercut*:

What is needed to retain the unqualified value of truth . . . is the possibility of
undercutting values or the absence of such, so that the complete absence of any
value beyond epistemic value may in some cases undercut the cognitive value in

9. Perhaps here I should make clear that I will not parley with those who would
argue that intrinsic values are incoherent—it is a presumption of this discussion that
entities of various kinds can be meaningfully said to have intrinsic value.

question . . . . In some cases, a neutral value for practical and other non-epistemic concerns leaves overall value intact (perhaps the deep truths about distant regions of our universe are like that) and in other cases a neutral value for practical and other non-epistemic concerns undercuts epistemic value, leaving the truth in question a pointless one. (204–205)

When the value of a true belief is undercut, then, this means simply that there is nothing else this belief has going for it aside from its truth—its value *qua* true belief is its *only* value. That does not eliminate the value this true belief has *qua* true belief; it is simply to grant, perhaps in concession to the likes of Sosa, that the *point* of *acquiring* a belief is always in reference to an interest that implicates (or appears to implicate) that belief. In this way Kvanvig claims that the value of a true belief can come apart from our interest in that belief. Does that not place him at odds with the basic dialectical strategy he seems happy to employ *a propos* of his claim that true belief is more valuable than empirically adequate belief? Not exactly.

For Kvanvig, the reason why particular true beliefs are intrinsically valuable has nothing to do with whether anyone finds them interesting. Rather, Kvanvig argues that true beliefs are intrinsically valuable as a class. Thus, individual true beliefs inherit value on account of their membership in the class of true beliefs. Why is the class of true beliefs intrinsically valuable? Because, Kvanvig insists, believing all truths is part of the "cognitive ideal":

> Part of the cognitive ideal, whatever else it may involve, is knowledge of all truths; omniscience, for short. But for omniscience to be part of the ideal, no truth can be pointless enough to play no role at all in the story of what it takes to be cognitively ideal. (209–210)

Why suppose that the cognitive ideal requires omniscience?

> Imagine a world with two beings, each claiming to be cognitively ideal. One is omniscient and the other is not. The less-than-omniscient being claims to be cognitively ideal in virtue of knowing all the important truths, but the omniscient being demurs. For among the important truths are the claims about what the omniscient being knows that the less-than-omniscient being does not know . . . . [T]he specific knowledge in question is also an important difference: that the omniscient being knows that the claim is true, for example, and that the less-than-omniscient being does not, establishes a significant different [sic] in terms of their grasp of the precise nature of the world in which they find themselves. (210)

Thus, for any given true belief that distinguishes these two beings, the fact that the omniscient being has this belief and the less-than-omniscient being lacks it is itself a significant truth. Since it is significant, each being must believe it. But if each being believes it, then presumably, each being believes the truth in question. (Though "X believes that p" does not follow from "X believes that Y believes that p" (where $X \neq Y$), it is reasonable that such an implication holds for beliefs about the beliefs of *omniscient* beings. Supposing omniscient beings have no false beliefs, learning that an omniscient being believes that p rationally requires believing that p.)

I am not convinced, however, that facts about the differences in beliefs between these two beings must be significant truths. Kvanvig merely claims that this is the case. As J. Adam Carter argues, one result of this is the collapse of the distinction between significant and pointless truths (2011, 290–291). Kvanvig skirts paradox—the very pointlessness of a truth is apparently what makes it interesting![10] But that is absurd. I find science fiction novels interesting and I find westerns uninteresting. That does not mean I must be interested to learn what differences between these genres explain my interest in science fiction and my indifference to westerns. Even if I were to become interested in this relational fact, that would not confer interest upon the relata. Similarly, just because one truth is interesting and another is not does not imply that the explanation for this difference is significant or interesting. And even if the fact of this difference was significant or interesting, that would not suffice to confer interest on the erstwhile pointless truth. Relata need not share properties of the relations they satisfy. Consequently, Kvanvig fails to establish the incoherence of a cognitive ideal that falls short of omniscience. Possibly, some true beliefs (e.g., the genuinely pointless ones) are not part of the cognitive ideal. No such true belief inherits the intrinsic value owed to those beliefs which comprise the cognitive ideal.

Given the failure of Kvanvig's attempt to demonstrate the value of *all* true beliefs, we should adopt the position that at least some true beliefs lack intrinsic value. This quickly presses us to ask by which standards are true beliefs to be measured for intrinsic value. The lack of an obvious metric is a strong motivating force for holding, with Kvanvig, that all true beliefs are intrinsically valuable (even if these values can be overridden or undercut). But such a metric is possible via the account of intrinsic value previously sketched, even if it ultimately fails to provide us with absolute certainty in judgments about the intrinsic value of true beliefs. Here, then, I would like to apply the inference to the best explanation argument for intrinsic value to the case of certain true beliefs, keeping in mind that more will be said toward the end of this chapter concerning the scope of this value.

10. *Cf.* Sorenson (2011).

Let us grant that we are interested in many true beliefs for their instrumental value. If I have a particular goal I would like to achieve, certain true
beliefs can help me to achieve this goal. But sometimes our goal is to acquire true beliefs, and without any expectation that these beliefs will serve
as instruments for further ends. Is it reasonable to assume that these true
beliefs are intrinsically valuable? According to Paul Horwich the answer is
yes—such an assumption is needed in order to "justify our pursuit of truth
in fields of inquiry such as ancient history, metaphysics, and esoteric areas of
mathematics—fields that may not be expected to have any pragmatic payoff"
(2006, 351). It is worth mentioning that we need not share Horwich's incli-
nation to discount the possibility of pragmatically justifying fields of inquiry
like mathematics and ancient history. The case of pure mathematics is an especially fascinating one. A pattern that has emerged throughout the history of
modern mathematics is the tendency for the natural sciences to appropriate
fruitfully theories and concepts of pure mathematics—theories and concepts
often devised by mathematicians with neither explicit nor implicit interests
in the physical application of mathematics. There is much truth to the idea
that yesterday's pure mathematics is today's applied mathematics, and that
today's pure mathematics is tomorrow's applied mathematics, etc. Although
this might appear puzzling to some[11]—how is it that mathematics pursued
for its own sake finds application in the physical world?—that would ignore
how natural science is deeply indebted to mathematics on a conceptual level.
The natural scientist is something of an opportunist when it comes to mathematics and modeling, freely taking from mathematics whichever concepts,
theories, or constructs seem helpful to her. Indeed, we should *expect* the natural scientist to profit from the rich and expanding conceptual space provided
by pure mathematics.[12] Similar defenses could be constructed for other apparently "useless" areas of inquiry, although the details will vary considerably
from case to case.

Nevertheless Horwich's point should be granted. After all, identifying
mathematical truths as (indirectly) instrumentally valuable is not a disproof
of their intrinsic value. While the instrumental value of pure mathematics
might help explain why governments fund research into pure mathematics,
that does not explain pure mathematicians' interest in pure mathematics. Nor
does it explain our more general abiding interest in getting to the truth about
topics even when we do not expect a payoff for doing so. Horwich ventures an
explanation for these behaviors:

11. E.g., Wigner (1960).
12. I owe this observation to Maddy, who argues further that the usefulness of pure
mathematics *depends* on its autonomy from the natural sciences. See Maddy (2007,
329–343).

It is presumably *because* most truths are useful in practical inference—and not merely to those individuals who *discovered* those truths, but also to all the rest of us to whom they are communicated—that our society, simplifying for the sake of effectiveness, inculcates a *general* concern for truth for its own sake. Of course, this causal/explanatory conjecture does not purport to explain the *fact* that truth is valuable for its own sake, but merely *our tendency to believe* that there is such a fact. The normative fact itself may well be epistemologically and explanatorily fundamental. (2006, 351)

Horwich may be right that the majority of our concern for the truth is regulative, but our concern for the truth explains more than merely our *tendency to believe* that truth is valuable for its own sake. Consider again the case of pure mathematics. No doubt most pure mathematicians find their specializations fascinating, are interested in getting to the truth about unsolved problems, are interested in getting to the truth about which statements are theorems of particular theories, etc. I assume that many mathematicians are interested in doing these things for their own sake. Valuing these pursuits for their own sake is part of the practice of pure mathematics, i.e., it is a useful component of the methodology of pure mathematics, and one that helps push the discipline forward. In this way the hypothesis that true mathematical beliefs are intrinsically valuable is indispensable for explaining this aspect of mathematical methodology. Thus, this assumption helps to provide a coherent, overall picture of mathematics—one that respects its centrality and importance to our overall scientific worldview. At such a juncture it is doubtful that any further or stronger evidence could possibly be provided for the intrinsic value of true mathematical beliefs. If that is right, and if we are entitled to assert of at least some things that they are intrinsically valuable, then we should endorse the intrinsic value of at least some true mathematical beliefs.

What I suggest, then, is that we apply similar reasoning to all legitimate practices within the scientific worldview. That is, within legitimate practices the interests of practitioners serve as evidence for which beliefs are intrinsically valuable. However, this quite clearly smacks of subjectivism and elitism. Why should the research interests of mathematics, scientists, etc., play a privileged role in the determination of which true beliefs are intrinsically valuable? And why should their interests provide grounds for *objective* claims about what is valuable? First, it is not my intention to restrict intrinsic value to true beliefs desired by practitioners. I agree with Stephen Grimm (2011a) that the class of legitimately interesting truths extends well beyond the confines of science proper. Second, I deny the charge of subjectivism. Judgments about what is interesting are purely subjective only if there are no intersubjective standards of evaluation. We find such a standard if we turn to virtue epistemology and virtue ethics.

According to Linda Zagzebski, our judgments about what is interesting, and what true beliefs are worth seeking (and thus count as intrinsically valuable on the view I defend) are constrained in a eudaimonistic way. In particular, the seeking of certain true beliefs is a *constituent* of the good life:

> I propose that the higher-order motive to have a good life includes the motive to have certain other motives, including the motive to value truth in certain domains. The higher-order motive motivates the agent to have the motives that are constituents of the moral and intellectual virtues, and in this way it connects the moral and intellectual virtues together . . . . [T]hat motive has nothing to do with epistemic value in particular; it is a component of the motive to live a good life. (2003, 24)

A Zagzebski-inspired justification for the intrinsic value of true belief is that having true beliefs of various sorts is partly *constitutive* of the good life. That is, having true beliefs is not a *means* to the good life but is *part of living* the good life. This raises a question about which true beliefs are parts of the good life. I will address this question in somewhat more detail toward the end of this chapter, after having considered whether our focus should not be on true belief but instead on either knowledge or understanding. But the essence of the response I would like to offer is that there are facts of the matter—contextual facts—that bear on which true beliefs (or items of knowledge, or understandings) a person ought to strive for. These would be facts about what is good for the kinds of beings we are, leading the kinds of lives we lead, in the circumstances we find ourselves in. All it takes for a true belief to be intrinsically valuable is for it to be a constituent of the good life of some individual—which ought to cover virtually every area of scientific inquiry, and more besides.

For the time being, however, I want to consider whether our focus should be on some other kind of epistemic state. It has been argued that *knowledge* is more valuable than true belief; as well, it has been argued that *understanding* is more valuable yet. So, I would like to investigate what should be said in favor of the intrinsic value of knowledge and of understanding. The latter notion is especially important in science, since scientific research is plausibly described as aiming not at the accumulation of mere true beliefs, but instead at *understanding*. As I shall argue over the next two sections, it is open whether knowledge and understanding have an *epistemic* value that true belief lacks. Thus, it may be that among these three epistemic states, only true belief is distinctively epistemically valuable—though as we shall see, understanding may manifest a new, non-epistemic source of intrinsic value.

Assuming, then, that some true beliefs are intrinsically valuable, it is natural to wonder whether epistemic states that include true beliefs, e.g., knowledge or understanding, are of any greater value. Intuitively, knowledge and understanding are more valuable than mere true belief: One seldom lauds the scientific enterprise for its assistance in our acquisition of true beliefs; rather, we laud the scientific enterprise for its assistance in the acquisition of *knowledge* and of *understanding*.[13] Similarly, given a choice between merely having a true belief about the way to Larissa and either knowing the way to Larissa or understanding how to get to Larissa, we would prefer the latter—and it seems we would prefer the latter even if, contrary to Socrates's view from the *Meno*, these options were practically or pragmatically equivalent.

Focusing for the moment on knowledge, the claim is that the state of "knowing that p" includes something which adds value above and beyond the value we would attribute directly to the state "truly believing that p." Consequently, one might require of an analysis of knowledge that it reveals the source of this increase in value. Since the publication of Edmund Gettier's (1963) well-known counterexamples to the "knowledge-as-justified-true-belief" thesis, epistemologists have attempted to identify some further criterion that when added to justified true belief avoids Gettier-style counterexamples and thus results in knowledge. The problem is that the various proposed criteria appear to mark knowledge only as more instrumentally (as opposed to intrinsically) valuable than justified true belief.[14] Consider, for instance, the process-reliabilist's proposal that knowledge is true belief that is justified via a reliable belief-forming process. The fact that a true belief is the product of a reliable belief-forming process seems in no way to add any intrinsic value to this true belief. However, a reliable belief-forming process would be instrumentally valuable, since those possessing reliable belief-forming processes would be more likely than others to acquire greater numbers of intrinsically valuable true beliefs.[15] The upshot is that, on the process-reliabilist's theory of knowledge, the intrinsic value of knowledge is nothing over and above the intrinsic value of true belief. This is one instance of what Kvanvig (2003) calls the "swamping problem"—in which the value of knowledge is "swamped" by the value of true belief. And, as Kvanvig has argued,

13. See Dellsén (2016) for an argument that scientific progress tracks increases in understanding rather than increases in knowledge.
14. Similarly, the value of *justification* in justified true belief seems only to be that justification is an instrument to acquiring true beliefs. See Kvanvig (2003, ch. 3).
15. The locus classicus of this critique of process-reliabilism is Zagzebski; see Zagzebski (1996) and Zagzebski (2003).

this problem is not unique to process-reliabilism but moreover arises for any account of knowledge that attempts to subvert Gettier-style counterexamples (such as Alvin Plantinga's (1993) "warrant," Robert Nozick's (1981) "truth-tracking," etc.). Therefore knowledge, without further comment, is not more intrinsically valuable than true belief.

As Duncan Pritchard has helpfully adumbrated, there are a variety of responses one might pursue when confronted with the swamping problem. One is what he calls the "practical response," which bites the bullet and accepts that knowledge is at best only more *instrumentally* valuable than true belief. This response aims to explain away, rather than codify, the intuition that knowledge is more intrinsically valuable than true belief (Pritchard 2010, 16–17). Second, there is what Pritchard calls the "monist response," which reverses the tables and maintains that knowledge is the sole intrinsically valuable epistemic state. On this picture, true belief derives its value from its relationship to knowledge, rather than the other way around (18–20). Finally, there is what Pritchard calls the "pluralist response," which denies that true belief is the only intrinsically valuable epistemic state. Should the pluralist fail to include knowledge among the intrinsically valuable epistemic states, she would, like the practical-respondent, owe us an explanation for why we had mistakenly come to believe that knowledge is more valuable than true belief (20–23). I agree with Pritchard that, on first inspection, the pluralist response is the most attractive option, i.e., that plausibly there is some epistemic state other than true belief that is of freestanding intrinsic value. However, as I shall argue, on closer inspection Pritchard's pluralist response is undermotivated and consequently, the practical response is a live option.

What epistemic state aside from knowledge might be thought to have freestanding intrinsic value? A number of epistemologists have argued that *understanding* is a kind of epistemic state that has value over and above the value of true belief. Though I will have more to say about the nature of understanding momentarily, for the moment we can take it to be (roughly) the grasping of connections among truths. Thus, to understand something—be it a proposition or theory—is not simply to have true beliefs (or knowledge), but to have a "grasp" on the various relations of dependence (inferential, explanatory, logical, etc.) among one's beliefs as they pertain to the proposition or theory in question. Below my focus will be on Pritchard's account of understanding, according to which understanding is a *cognitive achievement*. Cognitive achievements, he argues, are intrinsically valuable, and hence understanding is also intrinsically valuable *qua* cognitive achievement.[16] What I shall argue is that there is nothing uniquely *epistemic* about the value of cognitive achievements. Although I take this as an objection to Pritchard—who

16. Note that Pritchard prefers the term "final value," which for present purposes is no different from my notion of "intrinsic value."

claims that understanding is more *epistemically valuable* than true belief—this is nevertheless a welcome development in that there are coherent grounds on which to maintain that understanding is intrinsically valuable, even if only because it implicates true belief. Since understanding is often identified as an aim of scientific research, this provides an avenue for articulating the intrinsic value of scientific research.

While my focus is on Pritchard's account of understanding and its value (in part because his account is among the most thoroughly elaborated and discussed), I should mention to those familiar with the literature on the epistemology of understanding that I will not be detained by various debates that are, as it turns out, irrelevant to questions about the value of understanding, for instance: whether understanding is factive (i.e., requires only or mostly true beliefs); whether understanding is compatible with luck in ways that knowledge is not; whether understanding is always explanatory; whether understanding has a unique proper class of objects (i.e., propositions, or phenomena, or topics, or theories). Thus, that Pritchard's account of understanding is propositional, that it is compatible with certain forms of epistemic luck, etc., are largely irrelevant to issues pertaining to whether and how, on Pritchard's view, understanding is intrinsically valuable.

UNDERSTANDING AND COGNITIVE ACHIEVEMENT

With his account of understanding Pritchard intends to establish that the *epistemic* value of understanding is both intrinsically valuable and distinct from the value of true belief.[17] Something has *epistemic* value when it is either an epistemic goal or a means to an epistemic goal.[18] Thus, if knowledge is an epistemic goal and justification is a means to knowledge, then both knowledge and justification would have epistemic value. True belief is a good example of something that is both epistemically and intrinsically valuable, i.e., that has value not only within the confines of epistemology but also without restriction or without reference to any particular set of ends (perhaps save the end of living a good life). Meanwhile, what the swamping problem suggests

17. Moreover, in so far as there is a non-trivial overlap between understanding and knowledge on Pritchard's view, he claims that he can provide a diagnostic story about why we might have been inclined to think knowledge is intrinsically valuable—because knowledge is often compresent with understanding.

18. This is admittedly rough, since not just any precondition of reaching an epistemic end should count as an epistemically valuable *means* to an epistemic goal. Acquiring justification and acquiring coffee may each be preconditions of attaining knowledge (especially in mathematics!), but intuitively only the former has epistemic value, and a full theory of epistemic value would have to explain why this is the case. *Cf.* Bondy (2018) on the difficulties associated with defining "epistemic value."

is that, despite the intuition that knowledge is more valuable than true belief, analyses of knowledge imply that knowledge is merely an instrument for attaining the (epistemically and intrinsically valuable) goal of true belief.[19]

Pritchard's focus is on a variety of understanding known as "propositional understanding" or "understanding-why." This is the kind of understanding present when an agent has "understanding why such-and-such is the case" (Pritchard 2010, 74), e.g., "Mr. Humphries understands why Mrs. Slocombe brought her cat to work," or, "Linda and Eleni understand why there is at least a three-minute signal delay for one-way communications between Earth and Mars." Understanding-why stands in contrast to *objectual* understanding (also known as *holistic* understanding), which takes entities other than propositions as objects, for instance persons, topics, and theories. It is objectual understanding that is implicated in locutions such as "the Senator from Massachusetts understands the working class," "Professor Carolyn Porco understands orbital mechanics," and "No one understands continental philosophy."

For Pritchard, what distinguishes understanding-why from knowledge or true belief is that, when one has understanding, one has something more (or other) than a justified, Gettier-proof true belief. This something more is a grasp or account of how things "hang together" so as to bring about the understood proposition. Those who understand will normally have various context-sensitive powers which those lacking in understanding will not possess. Take, for instance, someone who merely knows or truly believes that there is a minimum three-minute signal delay for one-way Earth-Mars communications. Such a person could know or believe this without any conception of why this is the case, or of what in particular the signal delay depends on, or of how and why the signal delay varies over time, etc. Meanwhile, someone who genuinely *understands* why there is a minimum three-minute signal delay can accomplish at least some of these tasks. This highlights an important feature of understanding, *viz.*, that it admits of degrees. A person who understands why there is a minimum three-minute signal delay because she knows that communication signals travel at light speed and that Earth and Mars are never closer

---

19. It should be noted that Pritchard distinguishes between three different problems that he calls the Primary, Secondary, and Tertiary Value Problems (Pritchard 2010, 5–8). His Primary Value Problem is to account for the value of knowledge over true belief; his Secondary Value Problem is to account for the value of knowledge over that which falls short of knowledge; and his Tertiary Value Problem is to explain how the value of knowledge is different in kind from the value of true belief. *Kvanvig's* swamping problem, in turn, is a version of the Primary Value Problem that is restricted to *epistemic* value (15–16). Pritchard's focus, meanwhile, is the Tertiary Value Problem, which he claims addresses the intrinsic value (or what he would call the final value) of knowledge, and the solving of which he takes to provide concomitant solutions to the Primary and Secondary Value Problems (8). I shall treat the "swamping problem" as roughly synonymous with Pritchard's Tertiary Value Problem.

than three light minutes has some understanding of the signal delay. But a person who in addition knows or understands orbital mechanics has a deeper understanding yet, since she could account for how the signal delay varies on the bases of the orbital positions of Earth and Mars, how this delay might be different if Earth and Mars occupied even slightly different orbits, etc.

At this point one might be tempted to account for the value of understanding solely in virtue of the value of true belief. After all, what those with understanding appear to have is access to additional true beliefs or additional knowledge when compared with those lacking in understanding. But if Pritchard is correct, there is an additional element to understanding that we have not yet unveiled and which accounts for its *distinctive* intrinsic value. This additional element comes from Pritchard's analysis of understanding-why as a *cognitive achievement*—a cognitive success because of cognitive ability. Plausibly, the examples of understanding previously exemplified require varying degrees of cognitive or intellectual skill, and thus constitute corresponding cognitive or intellectual accomplishments. It is understanding-why's status as a cognitive achievement, then, that Pritchard takes to account for its intrinsic value.[20] His argument is straightforward: Understanding is always a cognitive achievement (80–83). Cognitive achievements are intrinsically valuable (66–73). Therefore, understanding is intrinsically valuable.

Pritchard motivates his second premise—the intrinsic value of cognitive achievements—by appealing to intuitions about the intrinsic value of achievements in general:

> Imagine, for example, that you are about to undertake a course of action designed to attain a certain outcome and that you are given the choice between merely being successful in what you set out to do, and being successful in such a way that you exhibit an achievement. Suppose further that it is stipulated in advance that there are no practical costs or benefits to choosing either way. Even so, wouldn't you prefer to exhibit an achievement? And wouldn't you be right to do so? (30)

He notes further that a good life is one that is "rich in achievement" (30–31), and thus, the value of achievements might also be thought of as constitutive of the value of human agency. Though I share Pritchard's intuition about desiring to exhibit achievements in such cases, I am not certain that this point alone

---

20. It is also the analysis of understanding as a cognitive achievement that distinguishes understanding from knowledge, according to Pritchard. For Pritchard, understanding is always compresent with cognitive achievement, while it is not the case that knowledge is always compresent with cognitive achievement. See Pritchard (2010, ch. 2) for his arguments, which also serve as objections to analyses of knowledge as a cognitive achievement.

establishes the intrinsic value of achievements, since I am not convinced our intuitions here are ultimately filtering away all practical considerations. The most plausible reason why someone would prefer to achieve rather than merely succeed is that, if they have the skills to achieve, they will be more likely to succeed in other cases. This concern is not decisive, if only because there is plenty of other evidence that exhibiting an achievement is something that we legitimately value for its own sake. Just think of the sundry of arguably prudentially valueless achievements—intellectual and other—that we are quick to value for themselves, admire, arrange competitions for, and reward: spelling, memorization, computation, plastic cup-stacking, frankfurter-eating, video game speed-runs, etc. A plausible hypothesis for the widespread phenomenon of holding competitions for apparently prudentially valueless skills is that we simply value achievements without restriction. If some or all of these acts of valuation occur in legitimate practices (in the sense described earlier in the chapter), then we have a defeasible reason for assuming that at least some achievements are intrinsically valuable.

However, even if they are intrinsically valuable, achievements are not by their nature *epistemically* valuable. So, if the value of understanding is owed to its status as an achievement, that would not establish that understanding is *epistemically* valuable. To see this, consider that one thing that we seem to admire or value about the individuals who participate in contests of various sorts is the sheer exercise of high levels of skill they often display. This admiration, respect, or value does not hold only of those who win their respective competitions, but rather for all of the competitors who demonstrate sufficiently high levels of skill (after all, the most skilled competitor does not always win the competition). In addition, we value the successes that these skills sometimes manifest, whether it is winning the competition, placing in the top ten, or simply succeeding in exercising a high degree of skill. In other words, it is plausible that we find both skills *and* successes intrinsically valuable, and do so regardless of their connection to epistemic ends. It is also the case that skills have an *instrumental* value in connection to successes—skills are instrumentally valuable on account of their ability to increase the likelihood of success. So when we speak of the value of achievements, we must be careful to distinguish between three possible senses in which we might say that achievements are valuable: The *intrinsic* value of the *success* involved in achievement; the *intrinsic* value of the *skills* involved; and the *instrumental* value of skill for bringing about success.

Note that Pritchard does not maintain that *all* achievements are intrinsically valuable. Rather, intrinsic value is restricted to what he calls *strong* achievements, which are "successes that are because of ability where the success in question either involves the overcoming of a significant obstacle or the exercise of a significant level of ability" (70). Strong achievements stand in contrast to *weak* achievements, which are simply "successes that are because of ability" (70). Though weak achievements need not lack value entirely, they nevertheless are never intrinsically valuable *qua* achievements (71). The distinction between weak and strong achievements is needed to account for our intuition that sometimes *easy* successes are intrinsically valuable, but only when they involve great skill (as when a skilled professional succeeds easily where those lacking the relevant skills could only succeed with great difficulty). Thus, for Pritchard, understanding is always a strong cognitive achievement, and understanding is intrinsically valuable because strong cognitive achievements are always intrinsically valuable.[21]

At this point we can appreciate Pritchard's solution to the swamping problem: Our mistake was in focusing on knowledge rather than understanding. While the value of knowledge may be swamped by the value of true belief, the value of understanding is not. This is because the distinctive intrinsic value of understanding compared with true belief is that only the former acquires value from being a strong cognitive achievement. Can the same be said regarding epistemic as opposed to intrinsic values? That is, does understanding, *qua* cognitive achievement, manifest a unique epistemic value? To see why Pritchard takes this to be the case, we need to acknowledge a further feature of his conception of understanding, which up to this point might appear to be purely internalist. To see why his is not an internalist account we need only recognize that understanding, as a strong cognitive achievement, does not depend "exclusively on the cognitive efforts of the agent" but has the external requirement that there be a "cognitive success" which involves "the right connection obtaining between cognitive ability and cognitive success" (82). Thus, arriving at the cognitive success of understanding-why requires not only coming to believe some true proposition, but also having a grasp on the *externally correct* dependence relations that account for the truth of the proposition (as opposed to having a grasp on some coherent-but-fictitious account of the truth of the proposition).

Pritchard's account of understanding is thereby a *factive* account of understanding. For instance, a phlogiston theorist might possess a coherent account of why some sample of wood burns at a certain temperature, and, by her own lights, she might claim to understand why this is so. Nevertheless, since she does not appeal to externally correct dependence relations (in this case, false causal or explanatory relations), she does not genuinely understand why the wood sample has the burning point that it does. Interestingly, Pritchard would not say that the phlogiston theorist understands nothing. He is willing to grant that she might have *objectual* or *holistic* understanding of phlogiston theory (for instance, she might have a grasp on the integrated set of dependence relations that comprise phlogiston theory).[22] This might even involve cases of genuine propositional understanding, e.g., understanding why such

21. See Carter and Gordon (2014) for an argument that understanding and strong cognitive achievements come apart.
22. Pritchard, personal communication.

and such is a prediction of phlogiston theory (perhaps together with the realization that although it is true that phlogiston theory makes such predictions, nevertheless these predictions are based on weak or unsound reasoning). Of this, more in a moment.

This external success requirement ensures that understanding is an *epistemic* state, and hence, that its intrinsic value is also an epistemic value. Presumably, then, it follows that its intrinsic epistemic value is not swamped by the intrinsic epistemic value of true belief. In reply I would like to motivate the worry that his external cognitive success requirement interferes with this result. Pritchard's account leaves open that the only *epistemic* value that understanding manifests is the epistemic value of true belief. This is because he does not establish that cognitive skills have any intrinsic *epistemic* value, though they may have whatever other intrinsic value that skills have more generally. Therefore, the only sense in which understanding could be said to be more intrinsically valuable than true belief is that understanding accrues value from the intrinsic value of the exercise of cognitive skills.

Problems for Pritchard

The central question I have for Pritchard's account is the following: In virtue of *what* is understanding an *epistemic* state? Given that it is a strong cognitive achievement, understanding has the following components:

1. It is a cognitive success.
2. It is attained through overcoming a significant obstacle or through exercising a significant level of ability.

Call (1) the "external cognitive success requirement" (or "success requirement" for short); and call (2) the "cognitive ability requirement" (or "ability requirement" for short). What I shall argue is that, between these two requirements, it is only the satisfaction of (1)—the success requirement—that provides understanding with epistemic value. And arguably, the epistemic value contributed by the success requirement is just the epistemic value of true belief. Thus, contrary to what Pritchard maintains (83–84), the *epistemic* value of understanding is swamped by the *epistemic* value of true belief. Showing this is a simple matter of showing that Pritchard denies understanding, and with it intrinsic and epistemic value, in cases in which (2) is present but not (1), i.e., in cases where only the ability requirement is satisfied.[23]

23. Such a response to Pritchard was anticipated but not developed in Janvid (2012, 194, n. 3). See also Ahlstrom-Vij (2013, 35–36) for a brief exposition of a similar argument.

Let me elaborate this objection in somewhat more detail before examining supporting cases: Assuming that Pritchard is correct that understanding (*qua* strong cognitive achievement) is intrinsically valuable, we can, in virtue of the previous discussion of the value of achievements, identify two sources of this intrinsic value:

3. The intrinsic value understanding has because of (2) the ability requirement, i.e., the intrinsic value it has *purely* as a display of cognitive skill; and,
4. the intrinsic value understanding has because of (1) the success requirement, i.e., the intrinsic value it has *purely* because it is a case of cognitive success.

In order for understanding to be intrinsically *epistemically* valuable, at least one of (3) or (4) must identify something of intrinsic epistemic value. Since, as I have already argued, achievements are *not* by their nature epistemically valuable, the only plausible source of intrinsic epistemic value here is (4). (It may of course be the case that *cognitive* skills are *instrumentally* epistemically valuable for bringing about cognitive successes, but that would still not establish that (3) is, in isolation from (4), of *intrinsic* epistemic value.) What I in turn shall argue is that, since Pritchard denies understanding (and hence intrinsic epistemic value) in cases where only (3) is present, i.e., where only the intrinsic value manifested by (2) the ability requirement is present, he fails to associate understanding with an intrinsic epistemic value that is distinct from the epistemic value of cognitive success, which is arguably just the epistemic value of true belief.

Consider again Pritchard's reaction to the phlogiston theorist's attempt to propositionally understand the burning point of a sample of wood. Though her mastering of phlogiston theory and her use of this to make a prediction about the case in question may satisfy the ability requirement—that is, while she may have overcome a significant obstacle or, what is more likely in her case, though she may have exercised a significant level of cognitive ability, nevertheless she does not satisfy the success requirement since what she grasps is an incorrect explanation of the burning point of the wood sample. This means that (2) and (1) can come apart, and likely come apart rather often, even when we acknowledge the implicit premise that (2) deals in particular with *intellectual* obstacles and abilities. Thus, if the ability requirement alone conferred intrinsic epistemic value, we ought to judge that this phlogiston theorist exhibits something of intrinsic epistemic value, even if she falls short of understanding-why.[24] However, unless we are willing to grant intrinsic *epistemic* value to *any* skillful exercise of cognitive effort, no matter how fictitious

24. Indeed, she exhibits what Grimm (2011b) calls "subjective understanding."

or irrelevant to the proposition in question, we should be wary of conceding that the phlogiston theorist, even though she fails to understand-why, nevertheless realizes something of intrinsic epistemic value.[25] So the denial of understanding in this case, and hence the denial of intrinsic epistemic value, hinges entirely on a failure to satisfy the success requirement; it matters naught whether intellectual obstacles have been overcome or whether intellectual abilities have been exhibited. Whatever intrinsic value we might nevertheless place on the exercise of these kinds of cognitive skills is, in consequence, non-epistemic.

As mentioned previously, there is *some* kind of understanding that the phlogiston theorist nevertheless might have, *viz.*, holistic or objectual understanding of phlogiston theory. This understanding consists primarily in the grasping of the dependence relations of phlogiston theory, perhaps just as an elaborate intellectual exercise. She might even have related propositional understanding. As mentioned before, she no doubt could understand why it is the case that "*p* is predicted by phlogiston theory" for various claims *p* predicted by phlogiston theory. Notice here that, in order to attribute understanding to our phlogiston theorist, it is again not enough that she merely satisfies (2), the ability requirement. Rather, on Pritchard's view, we have to invent some story about how she also manifests a cognitive success and so satisfies (1), the success requirement. Pritchard has (understandably) shifted the goalposts here: Rather than attributing directly to our phlogiston theorist the propositional understanding of the burning point of a sample of wood, we must retreat and claim instead that what she understands is phlogiston theory, or some proposition about the burning point being a prediction of phlogiston theory. This serves to exemplify again the idea that understanding-why (as well objectual or holistic understanding), and hence intrinsic epistemic value, is not manifested when one fails to satisfy the success requirement, and moreover that satisfying this requirement involves believing or accepting some true proposition or propositions. And so again, though our phlogiston theorist might exhibit intrinsically valuable cognitive skills, the exercise of these skills is not epistemically valuable.

This move, while perhaps palatable in the case of phlogiston theory, nevertheless cannot be maintained elsewhere without engendering considerable controversy. An important aspect of phlogiston-like examples is that they invoke evidently false dependence relations—not as potentially ineliminable idealizations as might occur in some scientific theories—but as patently false, scientifically facile assertions. Given our present understanding of combustion, there is no question that phlogiston theory is not a legitimate tool to use for advancing epistemic ends (at least as they concern understanding or

25. Janvid (2012) flirts with biting this bullet.

establishing the truth of claims about combustion). The problem is that there are areas of inquiry that, like phlogiston theory, are nomologically facile, but that nevertheless appear to be legitimate implements for reaching epistemic ends. Thus, the success requirement may go unsatisfied in cases where we have independently legitimate reasons for attributing understanding. This poses a difficulty under Pritchard's view about whether and/or how to account for the epistemic value of such (apparent) cases of understanding.

Consider, then, the various skills one might exhibit regarding a mathematical proposition, say, the Pythagorean theorem: by using it to calculate the dimensions of Euclidean triangles, by providing one or more proofs of it, by explaining its dependence on the parallel postulate, etc. I take it that our naïve, and not unjustified intuition is to assert without qualification that a person who could do all of these things *understands* the Pythagorean theorem and in so doing manifests something of epistemic value, if not intrinsic epistemic value. Indeed, this sounds like a textbook example of understanding-why (mathematical theorems are propositions, after all). However, on Pritchard's account, it is not clear that there is genuine understanding-why here, and so it is not clear that we can validate our intuition about the value of understanding the Pythagorean theorem. This is because the Pythagorean theorem is, strictly speaking, false: the parallel postulate on which it depends does not hold in our universe. Thus, this example does not satisfy the success requirement, since we are in violation of the factivity of understanding.

I do not see a way of responding to this case without adopting a controversial thesis in philosophy of mathematics—something that a theory of understanding should not be in the business of doing. Consider two approaches according to which our Euclidean geometer genuinely fails the success requirement and hence does not exhibit understanding. One approach is a crude version of Quine's naturalism about mathematics, which grants that unapplied mathematical theories are false, and in this way places these theories on a par, epistemically, with phlogiston theory. This would make our Euclidean geometer very much like our phlogiston theorist with respect to understanding-why. However, this kind of approach to naturalism, often because of its debasement of pure mathematics, has been widely rejected.[26] What about goalpost-shifting? We could deny understanding-why to our Euclidean geometer by claiming that all she has is holistic/objectual understanding of Euclidean geometry, or that all she has is some kind of indirect understanding-why: That all she understands is something like the fact that the Pythagorean theorem follows from the axioms and postulates of Euclidean geometry. There are at least two reasons why we should resist this kind of goalpost-shifting.

---

26. See Maddy (2005) for discussion.

First, Euclidean geometry is hardly unique in not mirroring the physical universe—we can expect that a great many other mathematical theories (along with their theorems) will also fail the factivity requirement. Consequently, we would be forced to revise the content of mathematical understanding-why on a massive scale. We would, in effect, have to adopt some form of fictionalism or if-thenism about mathematical assertions. While I happen to have naturalistically-inspired sympathies for such positions (Schwartz 2015a), a theory of understanding has no business forcing us to adopt such views. Second, and more importantly, goalpost-shifting only for unapplied mathematical theories invites the drawing of distinctions that are entirely irrelevant within mathematics itself. Euclidean and non-Euclidean geometries (including whichever geometry is "true" of our universe) are all equally legitimate areas of mathematical inquiry, and whatever account of propositional understanding we give in one case we should give in the other cases. Either all of mathematics resides within the epistemic domain, or none of it does.[27] Of course, one could "rescue" the example by insisting that mathematics, applied or otherwise, comprises a body of truths which do not require a revisionary semantics, but that would call for the adoption of something resembling Mark Balaguer's (2001) plenitudinous platonism, which is again something that a theory of understanding should not be in the business of doing.

I think it is not unreasonable to require that a theory of understanding *validates* our intuitions about understanding in mathematics, and it should do this even if the best philosophical account of mathematics holds that mathematics is not a body of truths.[28] If, for instance, Pritchard's account of understanding-why applies only to propositions that can be understood via non-mathematical (or, more generally, non-formal) means, then that would be a significant, unadvertised limitation of his view. (It would also be a highly implausible move, because a great deal of more "ordinary" understanding is mediated by mathematics.) A somewhat more attractive option would be to relax or allow exceptions to the success requirement in a way that allows for understanding-conferring successes in mathematics. For instance, for mathematical assertions, we might demand only the external requirement of logical consistency (and likewise for logical and other formal assertions).[29]

To avoid giving our phlogiston theorist a backdoor to understanding-why, we would need to construct a principled method for granting exceptions to

---

27. And, it would seem, a similar dilemma holds for logical theories and other formal theories.

28. See Maddy (2011, ch. 4) for a naturalistic argument that mathematics is not comprised of a body of truths.

29. Though admittedly there might be little distance between manifesting mathematical ability and manifesting mathematical success if the only requirement is logical consistency. This would trivialize mathematical success in a way that is likely to be objectionable to mathematicians.

the original success requirement. Why should we make an exception for mathematics, but not for phlogiston theory? This cannot be a merely ad hoc amendment. I do not mean to suggest that this cannot be addressed in a satisfying way—only that, insofar as Pritchard does not address this issue, his account of understanding is glaringly incomplete. Regardless of how Pritchard would choose to proceed—either by biting the bullet and denying understanding-why for large tracts of mathematical assertions, or by granting exceptions to the success requirement for mathematics, we again exemplify the idea that understanding-why, and hence its epistemic value, manifests only if a suitable success requirement has been satisfied. So, again, we are faced with the claim that satisfying only (2) the ability requirement is insufficient for manifesting understanding, and so satisfying only (2) does not manifest anything of epistemic value. Thus, the exercise of cognitive skill, including mathematical skill, is not inherently epistemically valuable (though it may often be *instrumentally* valuable in those cases where it helps to realize cognitive (and mathematical) success).

The upshot is that, on Pritchard's view, understanding, and intrinsic epistemic value along with it, manifests only when (1) the success requirement it satisfied alongside (2) the ability requirement, and not at all when only (2) the ability requirement is satisfied. This constitutes strong evidence that it is only the success requirement that contributes epistemic value to understanding, since the value of achievement (including cognitive achievement), though intrinsic, is not epistemic. Notice again that the kind of cognitive successes that occur when Pritchard grants understanding all involve true beliefs (e.g., understanding a true proposition by means of externally correct or valid reasons—consisting mostly of other true beliefs). Thus, if we want to know why it is that satisfying the success requirement manifests epistemic value (or, what comes to the same question, if we want to know why cognitive successes are epistemically valuable), we need only recognize that cognitive successes have an instrumental connection to true belief, or are simply identical to true beliefs.[30] Consequently, the epistemic value of cognitive success is either swamped by the epistemic value of true belief, or is identical to the epistemic value of true belief.[31] This complicates the sense in which understanding is intrinsically valuable according to Pritchard. As I see it, he faces a choice between two possibilities, neither of which establishes understanding as more *epistemically* valuable than true belief.

30. Or, in the case of mathematics, if our success requirement depends just on logical consistency, then the epistemic value of cognitive successes in mathematics would be in their instrumental connection to (or identification with) logical possibilities.

31. This point is not unique to the selection of true belief as the fundamental epistemic good served by (or identified with) cognitive success. Whatever else one might take as the fundamental epistemic good served by (or identified with) cognitive success, its value would still swamp (or be identical to) the value of cognitive success.

One possibility is that the intrinsic value of understanding comes *exclusively* from the intrinsic value of the exercise of cognitive skills. Since this intrinsic value is non-epistemic—i.e., since the only epistemic value of the exercise of cognitive skills is instrumental—the epistemic value of understanding would no longer be part of its *intrinsic* value. The epistemic value of understanding, then, would consist solely of the instrumental epistemic value of cognitive success. Thus, the epistemic value of understanding is swamped by the epistemic value of cognitive success, which I have just argued is swamped by (or identical to) the epistemic value of true belief. On the other hand, it could be that understanding derives at least some of its intrinsic value from the intrinsic value of cognitive success. This would require that any cognitive success implicated by understanding is intrinsically valuable. But again, as I have just argued, the intrinsic value of cognitive success, and hence the *intrinsic epistemic* value of cognitive success, is swamped by (or is identical to) the intrinsic epistemic value of true belief. And so again, the epistemic value of understanding is ultimately swamped by the epistemic value of true belief. In either case, of course, Pritchard is entitled to claim that understanding is more intrinsically valuable than true belief, because understanding manifests a value that truly believing does not—the intrinsic value of the exercise of cognitive skill. But to flog the long-dead horse, this value is non-epistemic.[32, 33]

Am I being unfair to Pritchard?[34] After all, it is certainly *possible* that some novel variety of intrinsic epistemic value is created through the admixture of cognitive ability and cognitive success that constitutes understanding. Even if I am right that the mere exercise of cognitive ability is not epistemically valuable, it could still be that the epistemic value of cognitive success does not fully explain the epistemic value of understanding. What is troubling here is that we only have Pritchard's word that understanding manifests a novel source of epistemic value. Perhaps it does, but we need some *reason* for thinking that this is the case. In the absence of such a reason—which Pritchard does not

32. *Cf.* Grimm (2012, 112): "[A]s a thesis about subjective understanding, Pritchard's view that understanding necessarily involves a cognitive achievement begins to seem much more plausible." What Grimm calls "subjective understanding" is basically the satisfaction of (2) with, at most, a merely *subjective* cognitive success. Grimm's observation, though correct in spirit, nevertheless needs to be articulated more carefully, since for Pritchard the obtaining of a genuine (non-subjective) cognitive success is necessary for cognitive achievement.

33. Two authors have provided an interesting diagnosis of what can go wrong under ability-based accounts of epistemic value. Melanson (2012) and Whiting (2012) each argue that views like Pritchard's conflate the value of *having* an epistemic state with the value of *arriving at* an epistemic state. My objection provides a way of vindicating their concerns: The intrinsic value of exercising cognitive skill is plausibly a value that attaches to *arriving at* understanding; meanwhile, the intrinsic value of cognitive success/true belief is plausibly the value that attaches to *having* understanding. See also Lynch (2017).

34. Thanks to Jeff Hershfield for bringing the following two concerns to my attention.

provide—we should be forgiven for attempting to explain the value of understanding in terms of the values of its parts. And when we turn to this task we find salient intuitions about the value of abilities and the value of successes—intuitions that make no essential reference to understanding. We place value on abilities (even when they do not manifest successes), and on successes (even when they are not the results of abilities). Pritchard has provided no argument for thinking that any value we place on understanding cannot be explained solely by appealing to these intuitions.

Another, deeper concern is that it might be erroneous in the first place to assume that "achievement" denotes a distinct ontological category. Stated somewhat metaphorically, perhaps "achievement" does not denote a natural kind, and there are only specific *types* of achievements, e.g., *athletic* achievements, *gustatory* achievements, *cognitive* achievements, etc., each possessing unique and incommensurate values. If that is right, then there would be no intrinsic value that something has simply by virtue of being an achievement, since there would be no such thing as an achievement *simpliciter*. This would undermine Pritchard's strategy for identifying the value of understanding, according to which understanding (*qua* cognitive achievement) derives its value from the value of achievements. While apparently devastating, a nearby solution is available: Appeal directly to intuitions about the value of *cognitive* achievements, which perhaps are intrinsically epistemically valuable in their own right. Indeed, I see no reason why we could not simply *initiate* discussion by acknowledging widespread intuitions about the value of *cognitive* achievements in particular. Nevertheless, we would still face a similar problem as before: that the intrinsic and epistemic value of cognitive achievement is explicable in terms of the intrinsic and epistemic values of its components. Thus, we would not be free of the need for an argument that cognitive achievements manifest unique or distinctive intrinsic or epistemic values.

Generalizing the Problem

Pritchard's account leaves open whether understanding manifests any distinctive intrinsic or epistemic value, and in particular, he does not show that understanding is more *epistemically* valuable than true belief. However, these lacunae are not unique to Pritchard's view. It is similarly open on several alternative accounts of understanding (including the views of Kvanvig (2003), Catherine Elgin (2017), and Henk de Regt (2017)) whether understanding has a distinctive intrinsic or epistemic value. This is because each of these views utilize their own ability and success requirements, leaving undefeated the possibility that whatever value understanding has can be fully explained by the values of its components. Since it would take us too far afield to discuss each of these views, I will pause only to rehearse my reasoning over Elgin's account of understanding.

Elgin's interest is in objectual (or holistic) understanding, and her analysis is far-reaching. In outline, she takes understanding to be

> an epistemic commitment to a comprehensive, systematically linked body of information that is grounded in fact, is duly responsive to reasons or evidence, and enables nontrivial inference, argument, and perhaps action regarding the topic the information pertains to. (2017, 44)

She places few limits on the kinds of representational devices through which understanding can be acquired or conveyed—she "does not privilege any sort of beliefs or representations a priori" but tolerates factive, verbal, pictorial, literal, figurative, denotative, and exemplificational representations (89). Though understanding must be "grounded in fact," it need not be factive, since she does not identify true belief or knowledge as requirements of understanding but simply that understanding's constituent beliefs are "acceptable." Whether an epistemic commitment is acceptable depends on whether that commitment stands in reflective equilibrium with other commitments. This process begins with some set of "initially tenable commitments" that "capture what we start out thinking we understand about" a domain and which "serve as touchstones against which to assess revisions" (66). From here we work to revise and improve upon this initial understanding using our most reliable methods. There is no particular end-state, but rather a gradual series of alterations in our commitments, each of which we regard as improvements over our previous commitments.

Despite its wider breadth and non-factivity, Elgin's and Pritchard's accounts of understanding exhibit a similar structure. On her account, understanding involves certain cognitive skills (the ability to use one's commitments to engage in nontrivial inference, for example) as well a criterion of success (that the components of understanding are acceptable). Similarly to Pritchard, Elgin is inclined to deny understanding in situations where skills are deployed in the absence of success:

> A coherent body of manifestly unfounded contentions does not constitute an understanding of the phenomena they purportedly bear on. Even if it is coherent, astrology affords no understanding of the cosmic order. (45)

And, again similarly to Pritchard, Elgin is happy to grant that there still might be *some* kind of success attributable to the astrologer, *viz.*, an understanding of astrology. But as she notes, "there is nothing epistemically special about this sort of understanding" (45). Nevertheless, she later discusses an example that might be thought to conflict with this judgment. In this example she asks us to imagine a group of scientists who unknowingly live in a Berkeleyan universe:

> What they take to be material objects are really immaterial ideas in the mind of God. Initially, they have fairly crude, unsystematic opinions about how things behave and about how to find out how they behave. Over time, they correct, extend, and systematize their approach to the point where they develop the scientific method. They subject their findings to peer review. They insist on controlled, repeatable experiments whose results are statistically significant. The geniuses among them come up with laws that reveal an astonishing order and regularity among the phenomena they observe . . . . Their account is wrong, however, in that it takes the substrate for that order to be material rather than mental. (85–86)

If truth were the standard of belief (and hence, of understanding), then we ought to judge that these scientists have achieved nothing of significance and that they do not understand anything about the astronomy of their universe. However, Elgin does not wish to deny understanding altogether to these scientists. Instead, she thinks that we ought to "acknowledge their significant epistemic achievement" (86):

> Otherwise we would have to construe the success of their science as a fluke . . . . Maybe someday they will be in a position to recognize that matter, as they conceive it, does not exist; maybe not. But it seems churlish to discount their achievement because it yields insight only into the structure of reality, not into its metaphysical ground. Rather, I contend, we should recognize that because their opinions, methods, criteria, and standards are in reflective equilibrium, they are reasonable in the epistemic circumstances. The system they have constructed is as good epistemologically as could have been achieved given their unfortunate starting points. (85–86)

Elgin's standard for belief—her conception of acceptance—is what permits her to attribute understanding in cases in which one believes falsely but nevertheless does "as good epistemologically" as one could have done. In the world she asks us to imagine, the scientists have commitments that they would deem acceptable, and this is enough for understanding. And so even in this case, the attaining of a "significant epistemic achievement" requires more than a mere adept deployment of cognitive skill—one's efforts still must result in relevant successes, even if the standard for success is acceptable belief as opposed to true belief.

To be sure, Elgin does not attempt an account of the value (epistemic, intrinsic, or other) of understanding. So it cannot be determined here whether she would endorse something like Pritchard's strategy for assigning value to understanding, or if she would prefer some other account. It is consequently open whether Elgin's account provides us with the means to argue that understanding manifests greater intrinsic value than any of the components of understanding. Nonetheless she sometimes speaks of understanding as constituting a "significant epistemic achievement" and certain failures as not "epistemically special." If we take these comments as proxies for claims about the presence or not of understanding's value, then my response to Pritchard is warranted: Since it is possible to account for the value of understanding through accounting for the value of its components—in Elgin's case, cognitive skills and acceptable belief—it is therefore open that the value of understanding is fully explicable in terms of the values of its components. While we cannot be certain that understanding, on Elgin's view, lacks a distinctive value, nevertheless no reason has been provided for thinking that understanding is distinctively valuable. If that is right, then whether or not an account of understanding is factive is irrelevant to its ability (or inability) to account for the distinctive value of understanding—since what appears important here is not whether understanding is factive, but instead whether understanding includes a success requirement.

No doubt these brief remarks regarding Elgin's accounts of understanding require a more substantial defense than I have provided here. Nevertheless, I hope I have shown that the features of Pritchard's account which are responsible for its leaving open whether understanding is distinctively valuable are not unique to his account of understanding. So perhaps this is a lacuna that appears generally under accounts of understanding that employ both success and ability requirements. Where this leaves us is somewhere short of a "pluralist" response to the swamping problem, since it has not been established that knowledge or understanding have an intrinsic epistemic value that is distinct from the intrinsic epistemic value of true belief. Nonetheless it still may be that understanding, at least on Pritchard's account, includes a kind of intrinsic value—the value of the exercise of cognitive skills—that is not implicated by true belief. Thus, there may be good reasons for preferring understanding to true belief over and above the reason that someone who understands likely has greater potential for acquiring further true beliefs. Nevertheless, this is compatible with the "practical" response to the swamping problem which holds that the only *epistemic* value of knowledge or understanding is the epistemic value of true belief (or acceptable belief *modulo* Elgin's account of understanding).

Admittedly we have drifted some distance from the central task of this chapter, which is to establish the intrinsic value of *scientific* beliefs, knowledge, and understandings. I think the detour has been a rewarding one, because we now have a deeper appreciation for what it is we value when we claim to value epistemic states such as knowledge and understanding. Thus, when we claim that scientific knowledge and understanding are valuable, this is an elliptical way of saying that we value true beliefs within the scientific domain, as well as the skills needed to organize these beliefs and acquire further ones. To close the chapter, I would like to see what further can be said about why such beliefs (and skills) are valuable, i.e., beyond the inference to the best explanation style argument I have used several times already.

VIRTUE ETHICS TO THE RESCUE?

It appears to be a working assumption of contemporary epistemology that there is a sharp and coherent divide between the epistemic and non-epistemic. Such an assumption has an important pragmatic value in that it helps epistemologists to prioritize the discussion and analysis of those concepts which are clearly epistemic in nature: true belief, justification, knowledge, etc. But as the previous discussion of understanding can (and perhaps should) be taken as evidence for, the divide between the epistemic and non-epistemic may be more blurry than sharp. Indeed, given the difficulties associated with identifying what is distinctively *epistemically* valuable about understanding, it may be that the best account of understanding is one which recognizes that it is only *partly* an epistemic state, and that its value is only *partly* epistemic (by way of the epistemic value of whichever cognitive success—true belief, acceptable belief, etc.—figures in understanding). Pritchard himself seems to recognize this implication about the value of understanding in a more recent paper:

> Understanding, for instance, is a good candidate for being an epistemic standing that has some broadly ethical value, and it will certainly often be practically valuable. Thus we shouldn't expect an account of epistemic value to *entirely* capture what is good about epistemic standings, but only specifically their *epistemic* goodness. (2016, 419)

This is all to suggest that the illumination we often seek—about the intrinsic value of epistemic states and values—true belief, justification, knowledge, and understanding—may not be attainable if we restrict ourselves to the conceptual resources of traditional epistemology.

As I argued at the beginning of the chapter, a possible way to account for the value of true belief (and with it understanding) is that the claim that certain true beliefs are intrinsically valuable is indispensable for articulating and explaining most, if not all practices that form a legitimate part of our overall scientific worldview. A similar conclusion seems warranted with respect to the intrinsic value of various cognitive (and other) skills, in so far as they also are implicated under legitimate practices. And, while I maintain that the scientific worldview stands in no need of a defense or justification from without, one might still be unmoved to accept the inference to the best explanation style argument I have offered for attributions of intrinsic value. Can I say anything more to recommend this dialectical strategy? I believe so, at least with some help from virtue ethics and virtue epistemology. On the one hand, valuing true beliefs is a requirement of virtuous conduct. If I care about something, then I must also care about getting to the truth about that thing. On the other hand, valuing the truth (or valuing cognitive success of some kind) is a component of wisdom and of a life lived well. That is, valuing the truth is *constitutive* of living the best possible human life.

The idea that virtuous conduct requires valuing the truth is best represented by Zagzebski. Actually, she goes further than this, arguing that "[t]here is no independent domain of epistemic value" because this value always derives from the values of other things we care about (Zagzebski 2004, 353). Her basic insight is that we have a conditional obligation to get to the truth about those things that we care about or place value on. Thus, if I care about the health of my newly planted azaleas, then it is incumbent upon me to conscientiously seek information about my azaleas and about what is good for them. This raises the question as to whether there is any *unconditional* obligation to conscientiously seek the truth. The existence of such an obligation would appear to require the existence of something we value or care about unconditionally. Zagzebski proposes morality as a domain that we (ought to) care about unconditionally, from which it follows that getting to the truth about morality is also unconditionally valuable:

> The importance of knowledge about moral matters puts a demand on each of us to be epistemically justified in those beliefs. There is a moral demand not to violate any epistemic demands. Given that there is an epistemic demand to be epistemically conscientious, there is a moral demand to be epistemically conscientious in my beliefs in the domain of morality. It is morally wrong to be epistemically unconscientious in any of these beliefs. Since these demands follow from the unconditional importance of morality, the demand to be epistemically conscientious in these beliefs and to conscientiously acquire beliefs in this domain is unconditional. (363)[35]

If she is right, then at the very least we have an obligation to care about getting to the truth about matters that pertain to morality. But can her perspective speak to the value of scientific research—especially space science—when

35. *Cf.* Baehr (2012), who argues for a similar dependence of epistemic values on moral values.

so much of this research seems rather far removed from ordinary ethical deliberation?

Zagzebski provides the basis for a response. First, part of being virtuous involves recognizing that the domain of information relevant to morality is not fixed and that the more knowledge one has, the better able one is to engage in moral deliberation:

> We are morally required to be epistemically conscientious in a very broad range of beliefs because of the social dimension of belief-formation. It is not at all obvious whether some belief will be relevant to moral judgment or action by myself or someone else who relies on my testimony, so that gives us a *prima facie* duty to be conscientious in a vast number of our beliefs. (367)

Thus, although a given domain of science may not appear relevant to morality *at present*, it may become relevant in the future. In the eighteenth century, theoretical physics might not have appeared in any way relevant to, e.g., human well-being, but looking back from the present we can appreciate how theoretical physics in the late nineteenth through mid-twentieth centuries affected human well-being through the development of concepts that could be applied in very helpful ways (nuclear magnetic resonance imaging; laser surgery) and very harmful ways (nuclear weapons). Second, it is also part of being virtuous that one helps oneself and others to acquire information about the domains they care about, which "broadens even further the range of beliefs that we are required to be epistemically justified in holding" (367). In so far as many scientists are not wrong in caring about the domains in which they research, we have (at least at a collective or societal level) an obligation to help see to it that they acquire further relevant beliefs in these domains. Therefore, on Zagzebski's account of virtue ethics and virtue epistemology, our unconditional concern for morality implicates a concern for getting to the truth about a wide variety of domains, including those domains that are of interest to scientific researchers, including space researchers. Thus, we are morally required to value the truth about these domains.

What's more, it is possible to argue for a stronger claim, *viz.*, that it is *part of living well* to value truth in these domains. A promising strategy here comes from Anne Baril, who argues that various beliefs and understandings, including scientific beliefs and understandings, are necessary for *human* flourishing. The reason for emphasizing *human* flourishing as opposed to flourishing *simpliciter* is that, according to Baril, we are not just any kind of epistemic agents; rather, we are specifically *human* epistemic agents, whose epistemic (and other) goals should be responsive to the wider goal of living a good *human* life. She insists that part of living a good human life is seeking certain *worthy* beliefs and *worthy* understandings:

To say that a belief is worthy implies certain norms, certain claims about what one ought to do. For example: to say that aesthetic insights and scientific understandings . . . are worthy implies that there is good reason to expose one-self to situations in which one might acquire such beliefs—to expose oneself to art, and to learn more about science. And this is true even if one doesn't care about art or science. Even if one doesn't care, one *should* care, since coming to have aesthetic and scientific insights (as well as caring about these things) is part of a good human life (whether one knows it or not). (Baril 2010, 229)

And, as she elaborates somewhat later,

[A] person who, for whatever reason, is deprived of *any* understanding of *any part* of the world we inhabit—the biological, the geological, the astronomical, etc.—is, in virtue of this deprivation, not capable of the fully flourishing human life that the rest of us are capable of. This is because, given the kinds of beings we are—rational, emotional, curious, etc., in the particular way that we are—such insights and understandings are part of living well as the kinds of beings we are. (236)

Baril's point, then, is that there are facts of the matter about what it means to live well as a human being at a given time and in a given situation. That this requires various *scientific* beliefs and understandings is compelling for us given that the scientific worldview undergirds the kinds of lives we live—at least for those of us living in societies organized around the scientific enterprise.[36] (And here it is worth reminding the reader that I unapologetically assume the legitimacy of the scientific worldview!)

We should endorse Baril's position; however, somewhat more needs to be said about the way in which certain beliefs and understandings become vital features of living well. For insight here I would like to turn to a somewhat un-expected source, though one to which I have already referred at several points, *viz.*, Penelope Maddy's naturalism—in particular her naturalism about math-ematical methodology. One of the most interesting features of the view she presents in Maddy (2011) is the radical proposal that the underlying reality of mathematics is neither a universe of mathematical abstracta (as might be favored by mathematical platonists), nor a class of modal truths about the implications of mathematical assumptions (as might be favored by mathemat-ical nominalists), but instead what she calls the *facts of mathematical depth*:

[T]his account of the objective underpinning of mathematics—the phenomenon of mathematical fruitfulness—is closer to the actual constraint experienced by

36. For related arguments see Neta (2008), Grimm (2011a) and Ryan (2012).

mathematicians than any sense of ontology, epistemology or semantics; what presents itself to them is the depth, the importance, the illumination provided by a given mathematical concept, theorem, or method. A mathematician may blanch and stammer, unsure of himself, when confronted with questions of truth and existence, but on judgments of mathematical importance and depth he brims with conviction. (116–117)

For Maddy, mathematics is an autonomous discipline, meaning that evidence for or against a mathematical claim must be all and only *mathematical* evidence. As she remarks here, fruitfulness plays an important evidential role in mathematics—whether mathematicians choose to adopt a new axiom, concept, or theory depends on whether adopting this axiom, concept, or theory would lead to tangible and recognizable progress in mathematics. What actually leads to progress is contextual, since it depends on the current state of mathematical understanding, the constraint of logical consistency, as well as the potential for the proposed theory or concept to answer existing questions and pose new questions of interest.

Meanwhile, from the point of view of logic, there is very little that distinguishes one consistent set of axioms from any other; nor is there much that distinguishes one well-defined concept from any other. Still, the selection of theories and concepts is not arbitrary, and can either open up or close off possibilities in logical space. Importantly, mathematicians are constrained in many ways when choosing a starting point in logical space for their inquiries: Their starting point must fall within the bounded space of logical possibilities provided by the content of current accepted mathematics. As currently practiced, some mathematical questions are of great interest; others are of little to no interest whatsoever. It is from this position that new concepts, axioms, and theories must demonstrate their relevance—either through the provision of solutions to existing problems, or by compelling practitioners to become interested in new problems. The facts of mathematical depth, then, are facts about which concepts, axioms, theories, etc., open up the right kinds of possibilities for mathematical progress, given the current state of mathematics and the requirement of logical consistency.

I cannot help but think the situation Maddy envisions for mathematics is closely analogous to our situation with respect to living well. Just as it is not open to mathematicians where or how to initiate their investigations, so too is it not open to us to choose a starting point with respect to what it takes to live well. We inherit the scientific worldview, and to repeat, there is no escape from Neurath's ship. We also inherit our humanity, and can do little to change that (at least for the moment . . .). Just as mathematical progress is constrained by logical consistency—since, given where we're at, only certain paths offer coherent solutions to existing problems—so too is living well constrained by a variety of modalities—logical, metaphysical, nomological, psychological—that place boundaries on which paths are open to us regarding our pursuit of living well. There are, I propose, "facts of eudaimonistic depth," which highlight the value of certain pursuits given our position in the space of nomological and human possibilities. These facts are not up to us, since they are merely relational facts about what pursuits would actually lead to human flourishing given our current status—although it is up to us whether we accept and act upon these facts.

From this perspective we begin with our existing understanding of science and of human beings, including what we know and understand about human flourishing. This includes the fact that we are human beings living in a modern, technological, scientific, democratic society. What holds for us might not hold for all humans, but we are not without justification in assenting to the superiority of the scientific worldview over other worldviews. What Baril, Zagzebski, and others have suggested is that having and valuing true beliefs, knowledge, and understanding (along with its component skills), especially in many scientific domains, is *constitutive* of living well, given our starting point is as already described. Thus, it is not merely part of the *best explanation* of practices in the scientific worldview that certain positive epistemic states are intrinsically valuable; it is also part of *living well* that one appreciates that these epistemic states are valuable and that one seeks to cultivate one's beliefs, knowledge, and understanding of morality, science, and other domains that exist coherently alongside the scientific enterprise—at least so much as one can, given the other demands of one's life situation. This, then, is the ground of the intrinsic value of scientific knowledge and understanding, at least under the scientific worldview (and I know of no worldview more worthy of adoption).

This also provides the grounds of an obligation to pursue scientific knowledge and understanding, because doing so would bring about things that have intrinsic value. This obligation is *prima facie* as opposed to being absolute or overriding, because the value of true belief, knowledge, and understanding are not absolute or overriding—I am not aware of many absolute or overriding values or obligations. Nevertheless, the obligation to pursue scientific knowledge and understanding is not altogether weak, at least compared to other obligations associated with true belief, knowledge, and understanding. This is because of the centrality of science in acquiring reliable information relevant to what we care about (be it morality or anything else) and to living well. Moreover, exceptional skill is on display in science—especially in space science, in which exploratory activity requires exceptionally high levels of skill and expertise. To play on Pritchard's terminology, the cognitive achievements of space science tend to be much stronger than the cognitive achievements of terrestrial science. Thus, the development and deployment of scientific understanding, especially within the space sciences, is nonpareil in its display of intrinsically valuable cognitive skills.

CHAPTER 3

The Instrumental Value of Scientific Knowledge and Understanding and the Rationale for Space Science

That scientific knowledge and understanding possess intrinsic value suffices to establish the pursuit of these things as *prima facie* duties. This is because of a general duty to bring about the good and because, other things being equal, good would be realized by engaging in activities that facilitate the growth of our collective understanding of the universe. I would even go so far as to say that the intrinsic values associated with scientific knowledge and understanding are significant in their extent, and consequently, that we have a strong, compelling obligation to engage in activities likely to expand on our understanding. Some, however, are skeptical of intrinsic value claims of the sort advocated in Chapter 2. Moreover, pointing to the intrinsic value of scientific understanding, as Kurt Bayertz (2006) has argued, is not especially persuasive in the public sphere as a rationale for science funding. We can, understandably, lament the disconnect between intrinsic values and government funding priorities. But we can also attempt to circumscribe more directly how the advancement of scientific understanding is instrumentally valuable for modern societies. It is much harder for the public to reject scientific research when that research is essential to the production of wide-ranging societal benefits.

In this chapter I argue that contemporary liberal democratic societies have an obligation to support a wide array of research projects. This obligation extends beyond characteristically "applied" forms of research (often valued for their apparent usefulness in the solution of social problems) to include "pure" or "basic" forms of research as well, including the space sciences. An important corollary is that scientific *exploration* is an especially valuable tool—if not an essential tool—for advancing science (even if there is no exact distinction between basic and applied research). The advancement of science, in turn, facilitates and contributes to democratic governance in various ways, for instance through cultivating and maintaining democratic culture. This instrumental, contributory value of scientific progress and scientific exploration highlights the value of much of space science, which by its nature places our scientific theories, ideas, and technologies in the kinds of novel circumstances that spur scientific change. As I argue in The Instrumental Value of Scientific Understanding, showing this requires more than the usual "spinoff" justification for space exploration. Undoubtedly space exploration has in the past produced many beneficial byproducts, but past performance is no guarantee of future success. A more substantive argument is needed, in particular, one which links scientific exploration, scientific progress, and social progress in a deeper, more secure way. Thus, I begin by relaying such an argument from Gonzalo Munévar, who holds that scientific exploration, scientific progress, and social progress are interdependent and interconnected. There is reason to believe, then, that scientific exploration (including space exploration) will continue to "spin off" many benefits to society.

Nevertheless, arguments like Munévar's might fall short of convincing those attuned to recent discussions in social epistemology concerning the democratization of science. Indeed, there has been a push—popularized in no small way by the work of Philip Kitcher (2001; 2011)—to reorganize science so that it more often and more effectively benefits the least fortunate. It is therefore worth showing that, even under democratically oriented accounts of the value of scientific research, there is an important role for extensive scientific research, including space research. To this end in The Warrant for Public Support of Science I construct a rationale for the state support of scientific research that is inspired by Mark Brown and David Guston's (2009) views on scientific freedom. The upshot of this is that the strength of the state's obligation to support a research project is a function of that project's potential to contribute to democratic deliberation. Happily, this implicates most forms of scientific research. As I argue, this implication holds even *modulo* Philip Kitcher's rather austere views on how democratic values ought to restrict research. This is because an important insight of Munévar's survives throughout this discussion: that exploratory research (even if it is not always "pure" or "basic") is instrumentally valuable for ensuring science's continued contributions to the democratic process, which is part of its value to society more generally. I follow this in Whither Space Science? with an overview of space research projects, identifying ways in which they fit into my overall rationale for the state support of research. This concludes my positive case for an obligation to engage in the scientific exploration of space. I close the chapter in Ethics *in* Space? with some brief remarks about how this obligation influences my subsequent discussions of planetary protection, space resource exploitation, and space settlement.

THE INSTRUMENTAL VALUE
OF SCIENTIFIC UNDERSTANDING

A common approach to justifying money spent on national space programs—NASA in particular—is to point to the many, varied, and *unanticipated* benefits of space exploration. NASA, for instance, maintains a website advertising the various technological and consumer "spinoffs" of work sponsored by the agency,[1] which include: tracking technologies used in laser eye surgery; scratch-resistant glass coating; the "space blanket" (also known as the "emergency blanket"); intumescent epoxy (a useful flame retardant); memory foam; a wax-based "Petroleum Remediation Product," NASTRAN (the NASA Structural Analysis Program); as well as many others.[2] The precise value of these spinoffs is difficult to measure, although an oft-repeated figure is that NASA has a return-on-investment ratio of 7:1.[3] It is tempting to argue for space research, as well as scientific research more generally, on roughly these grounds—that through the support of basic, pure, or exploratory research we inevitably will realize a sundry of unexpected benefits or payoffs.[4] However, some resistance to this idea, especially in application to space exploration, comes from Charles Cockell, who argues that the spinoff-style justification for space science

> is akin to saying that we should spend $25 billion building a giant model mackerel. In the process we will undoubtedly learn new construction methods; perhaps in the process of building scale models of mackerel we will learn new modelling techniques and we may even learn something unexpected about fish hydrodynamics . . . . In almost any project, whatever that may be, if you spend billions you will learn something new. (2007, 45–46)

In its place, Cockell urges, we instead should look for "very good *direct* reasons for going into space" (46). Though Cockell may be right that expensive projects of most any kind will provide learning experiences, it is not the

---

1. https://spinoff.nasa.gov/.
2. Meanwhile, Tang, Teflon, and Velcro were each developed *independently* of the space program, even though each is commonly identified as a NASA spinoff (likely because the space program made rather visible use of these products).
3. See Gurtuna (2013, ch. 4) for discussion.
4. The most well-known champion of this argument for basic research is Bush (1945). Below my focus will be on a kindred thinker, Munévar (1998), who provides a more substantive argument that is grounded in philosophy of science.

case that every such experience will be of equal quality and value. Building on the discussion from the previous chapter that some truths or topics are legitimately more interesting than others is the idea that some projects, in particular those which are legitimately scientifically interesting, are more likely than others to open up new strains of innovation in the generation of knowledge and technology. Thus, we may have good reasons to suppose that money spent on legitimately scientifically interesting research projects is likely to provide a higher "return on investment" than money spent on building a giant model mackerel. But can anything more substantive be said in defense of science and its instrumental role in the realization of societal benefits?

Gonzalo Munévar provides a compelling account of the "serendipity of science," i.e., the way in which scientific research is instrumentally valuable. His concern is with basic or pure research specifically, that is, research pursued for its own sake or without concern for applications (at least by those conducting the research). His perspective is rooted in the philosophies of science of Paul Feyerabend and Thomas Kuhn, who according to Munévar, insist that scientific theories are "instruments through which we conceive of the universe" (Munévar 1998, 174). As such, scientific theories are not mere devices of representation but also permit us to *interact* with the world through "seeing, hearing, analyzing, probing, and touching nature at many different energies and magnitudes" (174). Of critical importance here is that science often provides a plurality of interpretations of phenomena, and the differences between these interpretations allow for growth in our understanding of the universe:

> Different views of the universe, thus, lead to different assumptions, and different assumptions lead to different evaluations of what is to count as evidence. They also lead, as in the case of the Copernican revolution, to a profound transformation of our understanding of what the world is like. (175)

This has both theoretical and practical payoffs, thinks Munévar, because the way in which we understand the universe informs our sense of epistemic possibility. Differing scientific theories often implicate distinct conceptual resources as well as ontological and phenomenal categories. Problems and opportunities open according to one theory may be imperceptible on another. Not only has the progress of science (or "scientific change," to use Munévar's preferred terminology) enabled us to see more of the universe than before, it has also made us more aware of the dangers we face. As an example of the latter, little was known with confidence about the long-range consequences of runaway greenhouse effects until *Mariner 2*'s observations of Venus in 1962, which confirmed that Venus was subject to runaway greenhouse effects.

Not only does scientific progress (or change) influence our awareness of the problems we face and the opportunities open to us, it also affects the manner in which we devise solutions to these problems as well as the way in which we take up these opportunities. And, in many cases, scientific progress provides solutions to problems that simply could not have been anticipated by other means:

> Einstein began his career by asking "useless" theoretical questions such as "What would the universe look like if I were traveling on a light ray?" In trying to satisfy his curiosity about this and other equally impractical issues he was led eventually to develop his theory of relativity and to take a most decisive role in pushing physics toward quantum theory . . . . In these and other respects he changed several of our views of the world in profound ways, opening in the process the opportunity for a new understanding of light. This new understanding led to the theory of lasers. Lasers in turn opened up many technological opportunities. It was not long before some researcher decided to apply them in medicine. Today lasers are used in extremely delicate surgical procedures that would not be possible with any other technology known to medical practitioners. And it all began with a change of world view in a field far removed from medicine at the time. (175–176)

Another example of research in one area spurning technological advancements elsewhere is how research on the hydrodynamics of fish schools was employed in the design of wind energy systems. It was recognized that in certain situations, farms of densely packed vertical-axis wind turbines (VAWTs) would be more efficient than the much larger and more iconic horizontal-axis turbines (HAWTs), which have to be spaced at considerable distances from one another in order to avoid interference:

> [Researchers][5] noticed that the swimming animals did not need to stay far apart from each other. And they noticed something else about them: they formed a pattern of motion in which the swimmers did not all move in the same exact way, but whose motions complemented each other . . . . [T]he analogous approach on a wind farm would be to use VAWTs that were a good deal smaller than the large size that HAWTs needed to be to maximize their efficiency, locate these small VAWTs closer together, and let them take advantage of each other's wakes, by having each VAWT rotate in the direction opposite to its neighboring VAWT. (Sterrett 2014, 37)

5. *Viz.*, Whittlesey, Liska, and Dabiri (2010).

The more general point here is that technological and other advancements often are *precipitates* of progress in basic science. In much the way that pure, unapplied mathematical research finds its way into theoretical physics, so too does basic, unapplied research in science find its way—of necessity—into a sundry of applications.[6] In other words, basic research is something of a well that can be tapped for applications and for technology development. However, it is also a well that may dry up if it is not regularly deepened. For instance, it may be that our lack of success in finding cures for most cancers is because we are approaching the limit of the opportunities open to us under our current understanding of biology and chemistry. Our current situation with respect to cancer may be similar to that of doctors and scientists with respect to many illnesses prior to the acceptance of the germ theory of disease. Basic research of the right kind, e.g., a discovery of a second example of life on Mars, Europa, or Enceladus, could provide critical new insights into biology or chemistry which might help us to devise new and better cancer treatments (or even cures). Meanwhile, basic research of an entirely unexpected kind might well realize similar benefits.[7]

The final element of Munévar's position is that a significant antecedent of scientific progress (or change) is scientific *exploration*, i.e., the deliberate seeking out of novel and possibly anomalous observations:

> [S]cientific exploration places science in new circumstances and presents it with new ideas. Thus scientific exploration leads to not merely the addition of a few, or even many, interesting facts but to the transformation, perhaps the radical transformation, of our views of the world. (Munévar 1998, 176–177)

While it is true that scientific progress or change may be possible by looking inward (for instance, by discovering an inconsistency in one's theory), empirical adequacy—i.e., consistency with observation—remains an important value in the evaluation of scientific theories and hypotheses. However, our ability to determine whether a theory is empirically adequate is limited by the scope of our observations. Since it is always possible that new observations

---

6. Somewhat more substantively, this is partly a consequence of the widespread use in science of analogical reasoning. As research in one domain reveals new interesting patterns or structures, it is to be expected that scientists in other domains will recognize fruitful ways in which these patterns and structures shed light on the objects of their own research.

7. My claims about cancer research in particular are purely illustrative of a kind of possibility for basic biological research. I am not saying that current work on cancer is what Imre Lakatos would describe as a degenerative research programme—it might be, and it might not be. Nor am I saying that a discovery of a second origin of life would provide the kind of insights needed to improve upon current cancer research—it might do this, and it might not. My claim is merely that these kinds of discoveries have great potential for making contributions of the kind indicated.

might come to light which conflict with the predictions of our best theories, one of the best ways to test the empirical adequacy of our theories is to place them in novel or unfamiliar contexts. If the new observations conform to the predictions of our best theory, then we might increase our confidence in the truth (or at least in the empirical adequacy) of this theory. Meanwhile, if the new observations conflict with the theory, of if they turn up anomalies, then we must undertake a studious examination of our views. On the one hand, maybe the new observations were only *apparently* in conflict with or inexplicable under our best theory. But on the other hand, maybe the new observations suggest that we should amend, emend, or even reject and attempt to replace our current theory. In either case the result is meaningful scientific progress—progress made possible through the deliberate seeking out of novel information, i.e., through scientific exploration.

Putting the pieces of Munévar's position together provides the basis for an argument for the instrumental value of scientific understanding, as well as for the instrumental value of scientific exploration. The pursuit of scientific understanding (or knowledge or true belief) is instrumentally valuable because it leads naturally to serendipitous developments in other areas—developments that would not have transpired or in many cases have even been possible without substratal scientific research. Meanwhile, scientific exploration is instrumentally valuable because it is a highly effective adjuvant to the growth of scientific understanding. Thus, the instrumental justification for scientific understanding is one and the same as the instrumental justification for scientific exploration—that each is a means to realizing all of the various goods that humanity derives from the scientific enterprise.

A potential issue with Munévar's basic argument comes from its focus on pure or basic research—as opposed to applied research or technology development. The problem is that the distinction between pure and applied research may be more imaginary than real, as Heather Douglas argues while challenging the "linear model" according to which scientific progress begins with basic research that spurs applied research:

> [T]he pure vs. applied distinction is both artificial and implausible from the perspective of historical examination . . . . [T]he history of science reveals attempts at application as providing theoretical breakthroughs and theoretical work as providing new models of application. The linear model has been decisively rejected as descriptively inaccurate. (2014, 62)

No harm will come from accepting Douglas's observation that there is an interdependence between what are often described as examples of "basic" and "applied" research. Not only can apparently basic research suggest fruitful avenues for application (as Munévar argues), but so too can applied work suggest fruitful avenues for theoretical research. After all, sometimes when trying to solve a very local or technological problem one hits upon an idea of great theoretical significance. This means, however, that the idea of scientific progress cannot be conceptualized merely in terms of progress in theoretical or basic science but must apply to scientific activities more broadly. Douglas ventures a proposal, viz., that scientific progress be defined "in terms of the increased capacity to predict, control, manipulate, and intervene in various contexts" (62). "Theories or paradigms may come and go," she claims, "but the ability to intervene in the world, or at least predict it, has staying power" (62).

If we accept Douglas's view about the nature of scientific progress, then we ought to view the instrumental value of science in a more distributed way than Munévar encourages, since basic or theoretical work is not the only kind of research that facilities the prediction and control of our experiences of the world. Still, this does not detract from Munévar's claims about the value of scientific exploration. An important adjuvant to the progress of science, whether we think of science as pure, applied, or some nebulous admixture of the two, is placing our existing theories, concepts, technologies, instruments, tools, etc., in novel circumstances. It is through such exposures that we learn how to increase our powers of prediction, control, and manipulation. Whether our existing resources are successful or unsuccessful in these new circumstances provides an opportunity for progress: If they are successful, then we will have learned that our powers of prediction, control, and manipulation are greater than we had supposed them to be. If they are unsuccessful, then we will have learned something about the boundaries of our abilities, and in addition, we may uncover the kinds of anomalies that cause us to rethink our views, possibly at the theoretical level. So even if Munévar is wrong about the relationship between theoretical and applied work, there is still merit in the idea that scientific exploration spurs scientific progress, and is for that reason instrumentally valuable.

Furthermore, the instrumental value of scientific exploration remains even if we accept that the relationship between scientific exploration and scientific progress is bidirectional. Not only does scientific exploration spur scientific progress, but scientific progress also motivates and enables scientific exploration. Indeed, exploration is often required for the collection of observational data, which might in turn help resolve theoretical debates (as, for instance, astronomical observations are used in the adjudication of many debates in cosmology). Moreover, the results of scientific progress often suggest sites of interest for future exploration (as, for instance, developments in astronomy and cosmology spur the creation of new and improved instrumentation, which can then lead to progress in astronomy and cosmology, etc.). Recognizing that scientific exploration is an instrument of scientific progress does not require believing that scientific exploration must always occur prior to progress in related areas of science.

How does *space* science fit into this picture? Do we have a good reason to suppose that the many and varied space sciences conform to the general idea that scientific exploration facilitates scientific progress? I believe so, but I would like to hold off momentarily on developing this point. This is because, as yet, the instrumental justification for science is incomplete, since we have not shown that the instrumental justification for science is relevant to and compelling for those societies which might possibly engage in space exploration. For this reason, I will in the next section argue for public support of science (and scientific exploration) in modern, liberal democratic societies. In §3 I return to the topic of space science, where I explain how the space sciences factor in the case for public support of research.

THE WARRANT FOR PUBLIC SUPPORT OF SCIENCE

In part because space science is often derided as orthogonal to perceived societal goods and needs, it is worth investigating the grounds for the public support of science in a liberal democratic society. With any luck, the reasons why contemporary liberal democratic societies should support scientific research extend to most, if not all forms of space science—or at least, so I hope to make plausible by the end of this chapter. To this end I would like first to describe and refine a position according to which contemporary liberal democratic societies are obligated to support at least some forms of scientific research. For better or ill, most of the recent literature on the relationship between science and the state has focused on scientific freedom (or freedom of research), conceiving this as a kind of *negative* liberty: Scientific research is "free" when the state does not constrain researchers' choices in the selection of research topics and methods.[8] This would be what Torsten Wilholt calls scientific freedom of *ends* (2010, 175). My interest, however, bears more on what Wilholt labels scientific freedom of *means*, which refers to the idea of free access to those resources necessary for the conduct of scientific research.

A state that ensured scientific freedom of means would be maximally supportive of scientific research projects. Unfortunately, due to resource limitations, no state could be perfectly supportive of all research projects, or even of all worthwhile research projects. This means that states must make oftentimes difficult decisions about which projects are worth funding and which are not. Nevertheless, as I argue, what one says about scientific freedom of ends helps us to gain some traction on this latter issue: The research projects that

8. Or at least, such research is free when the state does not impose constraints beyond uncontroversial ethical requirements, e.g., that research involving human subjects is minimally harmful and is conducted with the voluntary, informed consent of subjects.

the state ought to support given its limited funds are a proper subset of those projects which the state ought to support were it endowed with unlimited funds. Meanwhile, those projects the state ought to support were it endowed with unlimited funds are a subset (and not necessarily a proper subset) of those projects that the state should not prohibit researchers from pursuing.[9] As I show later, there is an attractive rationale for scientific freedom of ends that transforms quite easily into an attractive rationale for scientific freedom of means. Therefore, in learning about the scope of scientific freedom of ends we learn something about the (somewhat more distant) boundary of scientific freedom of means.[10]

From Scientific Freedom of Ends to Means

What, then, of scientific freedom of ends? An attractive account comes from Mark Brown and David Guston (2009), who provide a defense of scientific freedom of ends which uses freedom of expression as an analogue. Within the United States, constitutional protection of free expression comes in degrees, with the strongest protections afforded to expression that either enables, informs, or is constitutive of the democratic process.[11] On this perspective, rights—including the right to free expression—acquire their force not as "natural, pre-political, individual protections against society and the state," but rather, "as a result of concrete struggles for social equality" (Brown and Guston 2009, 356–357). In other words, rights function to promote social equality, and in turn, to improve democratic governance. Thus, rights vary in strength based on their relationship to democratic governance, with those rights that are most closely connected to the political process accorded the highest strength. Following (Dahl 1985), Brown and Guston recognize three categories of rights, mentioned here in decreasing strength. First, there are "political rights" which "are integral to the democratic process (e.g., freedom of speech and assembly)" (Brown and Guston 2009, 360). Second, there are "social rights" which "are external to the democratic process but may be a pre-condition for it (e.g., basic education, health insurance, satisfaction of basic material needs)" (360). Finally, there are "civil rights" which "are external to the democratic process but entailed by human equality, and hence, supportive

---

9. Ignoring for the duration cases such as defense research which for relatively uncontroversial ethical reasons the state might support despite restricting the study of to state-sanctioned researchers.

10. I will not discuss the implications of this reasoning for *private sector* science, but I should not in any way be taken as suggesting that scientific freedom in the private sector is of little consequence. For discussion see Frankel (2009).

11. *Cf.* Weinstein (2009) and Post (2009).

of the type of moral culture that democracy requires (e.g., religious liberty, property, privacy)" (360–361).

Brown and Guston maintain that what holds for rights in general also holds in the case of a right to scientific freedom of ends (or what they call the "right to research" or the "right to inquiry"):

> [T]he right to inquiry is stronger when the inquiry makes a distinct contribution to democratic processes than when it does not . . . .
>
> From this perspective, inquiries that help resolve political questions would have the strongest claim to protection. This is not to say that other inquiries are without merit or should go without protection or support; rather, it is to acknowledge what the courts have tended to do with freedom of speech and the press anyway, that is, to create a hierarchy which, for example, prefers political speech and publication to commercial speech and publication, and both to pornography. (362)

Brown and Guston envision a hierarchy of forms of scientific research in accordance with their relevance to political, social, and civil matters. At the top, i.e., deserving of the strongest protection, are inquiries that help resolve political questions:

> There are many areas of inquiry that seem fundamental to the democratic process and should therefore be protected as such. Social-scientific knowledge may be used in the design of democratic procedures, for example, and natural scientific knowledge may provide vital information for effectively addressing social problems. (363)

Next are forms of research that "are not integral to the democratic process but contribute to its material preconditions (e.g., economic growth, scientific literacy, health, physical security)" (363). Last are forms of research that "support the democratic culture, broadly conceived, simply by satisfying intellectual curiosity or offering aesthetic pleasure" (363). Referring to these latter two forms of research, Brown and Guston claim that they

> deserve some degree of protection, but . . . they merit commensurably less protection than inquiries integral to the democratic process. Scientific research pursued primarily for curiosity's sake, for example, may deserve some degree of protection as a civil right, analogous to freedom of religion or the right to private property, but it may not deserve the stronger protection afforded to research conceived as a political or social right. Of course, given the unpredictability of science, the social and epistemic significance of research often changes over time, in which case its claim to protection changes as well. (363)[12]

12. As I shall touch on briefly in the Epilogue, *space* societies will experience a comparatively greater need to provide outlets for satisfying intellectual curiosity. So, it is

This raises the unenviable task of determining which forms of scientific research are protected to what degree—a task Brown and Guston do not attempt. There are, no doubt, an array of readily discernible answers: Climatology, medicine, and renewable energy research each cover projects that promise to help answer existing political questions and to contribute to the material preconditions of existing democracies. Meanwhile, virtually every area of scientific research helps to satisfy someone's intellectual curiosity. But these surface observations hardly help us assign precise degrees to the strength of protection that should be afforded to these various disciplines and projects. It would not be plausible to maintain that *all* climatology studies are protected, just as it would not be plausible to insist that all medical and renewable energy studies are protected. It simply isn't clear what granularity best suits this discussion.

Judging the merits of scientific disciplines is complicated by the difficulty of reliably and naturally demarcating disciplines, and by the fact that this level of analysis is far too general to distinguish between worthwhile and fruitless research projects, independent of their relationship to democratic deliberation. On the other extreme, judging the merits of individual experiments or theoretical postulates is hopelessly cumbersome, and prone to ignoring or trivializing the context in which the experiment or postulate would take place. I do not mean to suggest that the situation is altogether hopeless, just that there is much work to be done filling out the details of the basic framework Brown and Guston provide. I will engage in some construction in a moment, but I want to set this issue to one side so that I can connect Brown and Guston's discussion of scientific freedom of *ends* to an account of scientific freedom of *means*.

To transition from scientific freedom of ends to scientific freedom of means we only require the simple-minded assumption that a liberal democratic society has a duty to support those activities which are likely to contribute to democratic deliberation (*sensu* the political, social, and civil contexts adumbrated by Brown and Guston). Such a duty is *prime facie* rather than absolute, as the state has obligations other than those associated with the support of scientific research—duties which may in some cases conflict with, and override, its duty to support scientific research. Here, then, is the argument for state support of scientific freedom of means:

1. A liberal democratic society is obliged to engage in those activities which are likely to contribute to democratic deliberation (in the ways identified by Brown and Guston).

worth remarking here that it is a contextual matter whether "civil" forms of research warrant significantly less protection than "political" forms of research. Moreover, whether a given inquiry counts as political or civil also seems to be contextual. Researching the water content of an asteroid may for us be purely a matter of curiosity, but to an asteroid-based society it could be necessary for maintaining the material preconditions of the society.

2. Ensuring scientific freedom of means (and moreover ends) for at least some research projects is likely to contribute to democratic deliberation.

_____

3. Therefore, a liberal democratic society is obliged to ensure scientific freedom of means (and moreover ends) for at least some research projects.

This argument does not straightaway tell us which research projects a liberal democratic society is obligated to support—only that it must support at least some. Moreover, the strength of this obligation varies based on the manner and degree to which a research project contributes to democratic deliberation. Accordingly, the strongest obligations are tied to the support of inquiries which help resolve political questions, with weaker obligations tied to inquiries which merely contribute to the material preconditions of democracy or to democratic culture. Again, this leaves us quite in the dark about the details of the state's obligation to support scientific research. But at least we now know it has one![13]

Science: Rare, Medium, or Well-Ordered?

Can anything of greater substance be said about the kinds of research that the state ought to prioritize? That is, working within the rough framework provided by Brown and Guston, can we make concrete progress determining which research projects are most worthy of limited state research funds? Some help here is available from a relatively recent movement in philosophy of science which promulgates a more socially engaged conception of science and science policy. A central figure of this movement is Philip Kitcher with his notion of "well-ordered science" (2001; 2011). Kitcher bemoans the way in which, e.g., medical research seems focused more on profitable treatments and cures which benefit only the most fortunate, at the neglect of work that would be of significant benefit to the less fortunate, who are much greater in number. In particular, he sees this as a failure of science to respond to a genuinely representative assessment of the needs and preferences of society. This concern is a more specific version of a general call to socially responsible science, as Douglas provides with the entreaty that in order to construct an account of scientific progress that

_____

13. As Wilholt intimates, it becomes awkward to speak of scientific *freedom* of means when the means must be provided by the state, since "[i]n light of the limitation of resources, there is no such thing as a maximally inclusive right to the means for scientific research" (Wilholt 2010, 179).

sounds genuinely like progress, with all its positive connotations, we are going to have to embed science even more fully in society. We will need to ask which science will provide us with a better society, and, which science will perhaps undermine it. (2014, 63)[14]

To see what insights become available under this approach to science and scientific progress I will briefly lay out Kitcher's position, which on the surface is quite antithetical to more characteristically basic, theoretical, and exploratory research (even if, *modulo* Douglas, the distinction between theoretical and applied research is less than perfectly coherent). If there is an important role for these forms of research even on Kitcher's rather austere view, then we can be confident that the state ought to support basic, applied, and exploratory research on more relaxed accounts of socially, democratically responsible research.

Science is well-ordered, according to Kitcher, when it pursues those research projects that are identified through a process of ideal democratic deliberation as addressing *significant questions*. This deliberation involves neither the untutored preferences of the general public nor the autocratic ordinations of scientists but instead requires that ideal deliberators become informed about the facts and about research possibilities, with this information possibly transforming their values and their preferences. Meanwhile, those deliberators representing the scientific community must update their research preferences in light of the refined, tutored preferences of the other deliberators. In a nod to Rawls,[15] Kitcher expects these deliberations to be guided by the idea that science ought to emphasize lines of research that are most likely to benefit the least fortunate. A question is *significant*, then, when deliberators agree that it supports this kind of social good. Significance is a contextual matter, since it is responsive to current needs and the current potential of scientific research. Thus, the current significance of questions about climate change would be lost on those living two hundred years in the past, and (hopefully!) will be lost on those living two hundred years in the future.

Moreover, in the original articulation of his position, Kitcher sees significance as at least somewhat exclusive of purely epistemic considerations (such as those that often motivate basic or exploratory research):

[L]ess controversial than any duty to seek the truth is the duty to care for those whose lives already go less well and to protect them against foreseeable

14. See also Sarewitz et al. (2004, 77).
15. John Rawls (1971) is well known for promulgating the "maximin" principle— that an unequal distribution of society's goods is only preferable to an equal distribution when the unequal distribution makes the least fortunate better off than they would be under a system of equal distribution.

occurrences that would further decrease their well-being. We should recognize a clash of duties whose relative importance must be assessed. To oppose the argument, one must believe that the duty to seek the truth is so strong that it is binding, even in situations that will adversely affect the underprivileged, that will offer little prospect for gaining knowledge, and that will afford considerable opportunity for error. (2001, 103)

He is moreover not especially sanguine about the kind of defense of the value of basic and exploratory research offered by those like Munévar, who see this research as a conduit to societal benefits. Kitcher explains

The emphasis on the serendipity of discovery, which sees scientific breakthroughs coming in unanticipated ways, hardly demonstrates that what a group of scientists view as the hot topics for "basic science" will yield the means for satisfying the desires of a wider public. Strictly speaking, all that such examples show is that inquiry can lead to unexpected destinations. (141)

Kitcher, then, doubts the necessity and strength of the connection from progress in basic science to social progress, in no small part because he sees little progress on the kinds of problems that ideal deliberators would find most pressing, *viz.*, the problems faced by the poor and needy who make up the majority of the human population. Instead, Kitcher finds value in directed research, insisting that we "aren't completely clueless" when it comes to predicting the outcomes of research, because it is allegedly clear that

needs are more likely to be met if more effort is expended in certain lines of research rather than in others; nobody thinks that stepping up research into mechanisms of protein synthesis is likely to help solve the problem of global warming—it might but the probability isn't high . . . . We can only make the roughest of judgments, but we try to weigh the goals we have, think about available strategies in light of those rough judgments, and forge ahead. (2004, 56)

Wilholt is allied with Kitcher on this point, claiming that if the relevance of a research project "is truly unforeseeable in advance, then there is also no danger of targeted manipulation through meddling with prior research decisions" (Wilholt 2006, 259). If research has unforeseeable consequences, then it will continue to have unforeseeable consequences regardless of whether it is free or whether it is targeted or directed at particularly pressing problems. Bolstering this is Douglas's reminder that progress in science is a complex affair, sometimes stemming from practical work, other times stemming from theoretical work. Thus, serendipitous consequences are not exclusive to free basic or free theoretical research.

Admitting this, however, does not constitute an argument for the null- or disvalue either of basic research or of exploratory research. To see this, we need to distinguish between two ways in which research can be said to be serendipitous. Though scientific theories need not be closed under deduction, and although it is not descriptively accurate to characterize applied research as merely carrying out deductions from theory, there is nevertheless truth to the notion that current accepted theories place boundaries on the space of possibilities open to researchers of all stripes. Serendipitous consequences of applied research, then, are mostly beneficial in helping us to recognize possibilities and consequences open to us already, but that for whatever reason we had not before recognized. In other words, practical or applied research often helps us to *realize the potential* of our existing theories, powers of observation, powers of prediction, and powers of control. Meanwhile, progress in basic or theoretical work is serendipitous in the sense that it *expands the potential* of our scientific theories, powers of observation, powers of prediction, and powers of control. What this suggests is that Kitcher's notion of significance is somewhat impoverished, since it emphasizes research that might help realize existing potentials, whereas it discounts research that might help expand our potential powers. Exploratory research, since it can take place at any point in the applied-basic spectrum, is thereby serendipitous in each of these ways.

Matthew Brown (2010) recognizes something like this deficiency in Kitcher's account of significance, arguing there ought to be greater room for a more characteristically epistemic conception of scientific significance. In particular, he seeks to expand the notion of significance to include what he calls "genuine problems." For Brown, a genuine problem arises whenever there is a disturbance in a practice (e.g., a realization that a theory is inconsistent; or the discovery of data in conflict with prediction), or what he (following John Dewey) calls a "problematic situation." He takes "real" or "living" doubt about a practice to be a precondition of these disturbances. Idle speculation or Cartesian doubt is insufficient here:

> [D]oubt arises in the course of activity, and presents a blockage of such activity . . . . Doubt is a hesitation in the vital beliefs that structure that activity and that press for solution. Doubt is a discoordination that is not merely a subjective feeling . . . . The mere questioning of a thing does not make that question significant in any degree. (143)

Brown accepts that not all practices are equally important (at least in Kitcher's sense of importance). However, he claims that the significance of a problem is contextual not only in the way Kitcher imagines—as a function of the societal importance of the practices it impacts—but also in in the extent to which it impacts those practices:

We can imagine a small disturbance in a quite important practice may be very important. For example, suppose that we become aware of even a relatively small flaw in the practice of vaccination, such as a very low level uncertainty about its side effects. Because of the importance of vaccinations to modern medicine, this presents itself as a crucial matter. Second, consider a rather large disturbance in a much less important practice. Suppose you put very little stock in research in higher energy physics. Nevertheless, a problem which shakes that area at a fundamental level might be quite significant indeed. (145–146)

While I agree with Brown that we need to recognize the "individual, particular, and situational" nature of the scenarios which give rise to genuine problems (143), and to acknowledge the potential for basic research to count as genuinely significant, it is again worth remarking upon the role of exploratory activities. Those disturbances in practices which generate problematic situations, even within highly contextualized settings (be they in applied or basic domains of research), do not appear out of nowhere. Often enough these disturbances are the result of placing scientific theories, ideas, and tools in novel situations. Exploration, in this sense, is a generator of problematic situations, and thus of significant areas of research.

Kitcher has in his more recent work reserved a conditional position for basic or theoretical research in his well-ordered society. Although he remains doubtful about the frequency at which undirected basic or theoretical research has led to social progress of the kind conducive to a well-ordered society,[16] he nevertheless maintains that information about the potential benefits of basic or theoretical research, even if they are indirect, "*should* be part of what the ideal discussants know" (Kitcher 2011, 120). Thus, in Kitcher's hypothetical ideal democratic deliberation, deliberators are obliged to consider the serendipity of research in their determination of which questions and problems are truly significant, so long as these deliberations are based on "genuine knowledge about social direction of inquiry, success of brilliant individuals, or fruits of research into pure topics" (119–120). Though he is ultimately uncertain that "pure questions" would figure in well-ordered science, he nonetheless suspects that his ideal discussants would reserve a role for basic or theoretical research:

Perhaps because the knowledge available is, in some areas, so powerful, so susceptible of further development, and precisely because it has often grown out of programs of "basic" research, it appears highly unlikely that the ideal deliberation would abandon so profitable a historical strategy . . . . Unless you suppose

16. And here it is worth acknowledging Simon (2006)'s criticism that Kitcher's well-ordered science seems impossible without a well-ordered *society* that is capable of directing and conducting this science.

the situation is truly critical, that our species faces practical problems that command direct attention, well-ordered science is likely to maintain a role for "basic" research. (123–124)

Kitcher, then, is at least somewhat amenable to the value, within his well-ordered society, of more characteristically basic or theoretical research. I think for this reason he also ought to value exploratory research, since this work is intimately connected to the generation of novel insights which can be put to profitable use.

The upshot is that on Kitcher's rather restricted conception of the value of scientific research there are good reasons for preserving a role for the state support of basic research and scientific exploration. This provides us with a strong case for the state support of these forms of research, since the main resistance to them under Kitcher's view (and under other socially oriented accounts of scientific value or significance) is that undirected basic and exploratory inquiry is (purportedly) insufficiently instrumental to the realization of societal goods. If even Kitcher must recognize tangible value in these kinds of pursuits, then alternative positions that place a higher premium on epistemic values carry with them even stronger and more expansive mandates to support scientific exploration and basic research. Consequently, contemporary liberal democratic societies (even those especially interested in improving the lives of the least fortunate) have compelling reasons for supporting a broad array of scientific inquiries, including at least some inquiries which on the surface appear of purely epistemic interest.

WHITHER SPACE SCIENCE?

There are, then, compelling grounds for wide-ranging state support of scientific research, including basic and exploratory research. As I shall explain here, this extends to space research, resulting in an obligation to support various forms of space science and exploration. However, in the absence of a full overview of all possible research projects across all disciplines, and moreover in the absence of a more general evaluation of science funding compared to other areas of government spending, it is difficult to speak in absolute terms about the strength of the state's obligation to support space research in particular. It is possible that space research, compared to other forms of research, is less likely to realize the kinds of societal-democratic goods mentioned by Brown and Guston, among others. Thus, holding science budgets fixed, it may be that we are not presently in a position in which there are compelling reasons for increasing space exploration budgets. It is also possible that space research is worthy of increased support, even if overall science funding remains level. I do not intend to explore whether the former or latter (or some other possibility) is more likely. I am, of course, of the opinion that not enough is spent on science across the board; and that the United States would probably be better served if its government were to spend less on defense and more elsewhere, including more on scientific research and on space research. Still, I am content to show here merely that space exploration does indeed possess instrumental value for liberal democratic societies; and showing this does not require that space exploration is in this manner nonpareil. It simply requires that space exploration, as a rule, satisfies various rough criteria such as: helping to answer political questions; contributing to the material preconditions of democracy; and contributing to democratic culture. There is clear evidence that many space exploration activities fall under each of these headings. Nevertheless, even if we decide there is sufficient warrant for increasing spending on space research, the great costs associated with spaceflight will still limit the number and diversity of research projects. We would do well to keep in mind Wilholt's observation that space exploration "is doomed to be planned science due to limited resources" (2006, 262); there will never be perfect freedom of means for space research.

To begin with a general claim in support of the instrumental value of *space* science, Munévar notes that the exploration of the solar system is especially likely to turn up anomalies and to generate the conditions associated with scientific change (which in turn facilitates scientific and societal progress). Thus, according to Munévar,

> the reason we can expect a bounty from space is precisely that the exploration of space presents many challenges to our science and our technology. Since space exploration is thus so likely to contribute to the transformation of our views, investing in it has a clear advantage over investing in fields not so ripe with challenge. (1998, 177)

I am not certain that space exploration is evidently superlative with respect to its challenges. The interplay between instrumentation, experimentation, and theoretical work in fundamental physics presents many challenges, as does our various attempts to generate understanding about the deep sea and about Earth's climate. Still, Munévar correctly indicates that space exploration presents the kinds of challenges that spur scientific progress—challenges which test our powers of observation, prediction, and control; and which call for the mutually supportive development of new ideas, concepts, and instrumentation. Thus, space exploration has the potential to contribute to democratic society in all of the ways Brown and Guston mention. However, it remains to be seen which space research projects fall under which of Brown and Guston's broad headings (political, social, civil). Recall that, on the Brown-Guston framework, the strongest obligations are associated with research that addresses political questions, with commensurably weaker obligations associated with research that addresses social and civil issues. I will now provide overviews of which forms of space research fall under these broad headings, which provides a provisional means for assessing the relative strength of our various obligations to support space research.

Research That Helps Resolve Political Questions

One of the most politically potent forms of space research is Earth observation from space. From various orbits around the planet, satellites have the capability to provide us with an extraordinary amount of policy-relevant information about weather, climate, resources, development, etc. Space-based remote sensing of Earth takes place across the electromagnetic spectrum. Optical sensing, which operates in the wavelength range of 0.3 to 14 μm (covering the visible spectrum of 0.4 to 0.75 μm), makes use of digital imaging, radiometers, spectrometers, as well as laser and radar systems. Optical sensing is useful for measuring surface temperature, surface vegetation, and for evaluating crop health and making crop yield predictions (Dech 2006, 56–57). Radar systems are especially valuable here because atmospheric haze and cloud cover often interfere with optical imaging. Another form of remote sensing is synthetic aperture radar observation, which operates within the microwave range (3 to 25 cm) and can penetrate vegetation, soil, and snow, and which works regardless of weather or illumination conditions. This allows for measurements of ground motions, including motions associated with landslides, glaciers, volcanoes, and land subsidence (58). Optical sensing, then, is especially important to agricultural and forestry studies, while synthetic aperture radar provides information about Earth's topography, earthquakes and volcanic eruptions, and ocean currents. Space-based remote sensing also provides observations of Earth's weather as well as data about changes in Earth's atmospheric composition and about the interactions between the oceans and the air. Simply put, our knowledge about Earth would be much impoverished without access to space-based remote sensing.

All of the above information is relevant to a sundry of political issues ranging from agricultural policy to resource exploitation decisions to disaster prediction, warning, and response. However, studying Earth itself is not the only way to acquire knowledge and understanding relevant to the resolution of political questions. For instance, by engaging in comparative planetology, i.e., by studying the differences between the planets, we can come to better understand atmospheric and geological processes on a general level, which can enable the generation of new insights that might assist in our attempts to mitigate the effects of anthropogenic climate change. As one example, Olivier Mousis et al. have argued that in situ examination of Saturn's atmosphere promises rewards on two fronts: First, it will increase our understanding of the formation of the Solar System; and second, it will improve our understanding of atmospheric processes:

> Both themes have relevance far beyond the leap in understanding gained about an individual giant planet: the stochastic and positional variances produced within the solar nebula, the depth of the zonal winds, the propagation of atmospheric waves, the formation of clouds and hazes and disequilibrium processes of photochemistry and vertical mixing are common to all planetary atmospheres, from terrestrial planets to gas and ice giants and from brown dwarfs to hot exoplanets. (2014, 30)

Thus, what we learn as we explore other planets can help us to understand terrestrial processes. This is significant because we suffer a version of the "$n = 1$" problem, in which we cannot say which features and processes of Earth's atmosphere are contingent, which are shared by other (or all) atmospheres, and which are unique to Earth. A more general understanding of atmospheric processes (facilitated by comparative planetology) would help us to generate research proposals that might reveal answers to some of our most pressing inquiries about Earth's atmosphere and climate.

However, we should be careful not to oversell the value of comparative atmospheric studies. While a serendipitous solution to problems related to anthropogenic climate change may result from increased study of atmospheric processes on other planets, such insights may not become available on short enough timescales. The orbital characteristics of the planets significantly limit the frequency at which missions can be sent out. We cannot simply send a scientific payload to the gas giants Jupiter or Saturn whenever we please; rather, missions to these worlds are typically launched only when the relative positions of the planets allow for low-energy transfers from Earth. To save on fuel, missions to destinations beyond Mars often employ gravity-assist maneuvers. The *Galileo* mission to Jupiter made use of a craft launched aboard the Space Shuttle *Atlantis* in 1989. Following gravity assists both from Venus and from Earth, it entered Jupiter's sphere of influence in late 1995, where it gathered data until 2003 when it was deliberately crashed into Jupiter's atmosphere.[17] Therefore the time-horizon for data retrieval is on the order of years to decades from the point of launch, and generally on the order of decades from the point of mission proposal. (For example, the James Webb Space Telescope, which has been in development since 1996, is currently awaiting a launch in 2021, provided there are no further delays.) This may be too long of a time horizon for effective data acquisition, analysis,

---

17. Had it been left in orbit, the spacecraft might eventually have impacted on Europa. Since it had not been sterilized, it could have contaminated this moon, which has a subsurface ocean that potentially harbors life.

incorporation into theory, and technological application, especially if the goal is to use missions to other planets to help solve environmental problems on Earth. We should not, then, take the promises of comparative planetology to absolve us of a continuing obligation to grow our existing understanding of terrestrial processes, and to use our existing understanding, to the best of our abilities, to address the ecological crises we face. Comparative planetology, while it may help us to better understand Earth over the long run, cannot be substituted for terrestrial environmental research. Planetary research and terrestrial research are *complementary*.

Research That Contributes to the Material Preconditions of Democracy

Recall that according to Brown and Guston, research contributes to the material preconditions of democracy when it facilitates economic growth, scientific literacy, human health, and national security. In Chapter 1 I outlined how some of these considerations are part of the widely promulgated "space advocacy package" of rationales for space exploration. There I noted the paucity of evidence of spaceflight's impact on scientific literacy, and I also expressed some provisional doubts about spaceflight's ability to promote economic growth. Regarding the latter of these, it is worth specifying that one particular concern I raised, which I will take up in more detail in Chapter 5, is that it is unclear to what degree the majority of humanity would benefit from the commercial exploitation of space resources (such as lunar and asteroid materials). So what I shall consider here are the potential benefits—economic and physiological—of more near-term space research.[18] Interestingly, Wilholt seems especially skeptical about this kind of value of space research, stating that it does not seem to him that "mapping the stars" has "a discernible relevance for the democratic process" (2006, 259). Astrometry is used in precision time keeping, so it is not without economic merit, though it is perhaps less important outside of the sciences than it was when astrogation was a principal method for ship navigation. Still, simply because one area of space science is of limited relevance to economic growth and human health does not mean the same holds for other areas of space research.

The study of living organisms in reduced- and micro-gravity can improve our understanding of biology and human physiology in a number of ways. As Ian Crawford relays, among the life-sciences opportunities available on space missions—especially those with human crews—are studies on the effects

---

18. I am not much interested in issues pertaining to the use of space for national security, though I point the interested reader to Schrogl et al. (2015) for an overview of issues related to space power and the military uses of space.

of reduced- and micro-gravity on gene expression, immunological function, bone physiology, and neurovestribular and cardiovascular function (2005, 247). We do not know, for instance, whether there are "gravity thresholds" for various aspects of human physiology. While this kind of understanding would be directly relevant for those seeking to determine whether human space settlement is viable, it also promises to shed new light on issues faced by terrestrial humans, such as osteoporosis, muscle atrophy, cardiac impairment, and balance and coordination defects (247).[19] Meanwhile, there are many opportunities for materials research in space. The design of materials to withstand exposure to vacuum and to various other environmental conditions in space can lead to economic opportunities on Earth (think again about the "space blanket"). Moreover, little is known about the effects of gravity on crystal formation. Studying crystal formation in a microgravity environment, such as in Earth orbit, would allow for "the study of crystallization or solidification of fluids not disturbed by gravity induced fluid flow or, alternatively, allows the study of the effect of flow under control of the scientists, not disturbed by 'natural' or gravity dependent convection" (Ratke 2006, 300). The presence of gravity also prevents precise measurements of various thermophysical properties of materials, such as heat capacity, surface tension, and electrical conductivity, since the presence of gravity disturbs convection processes (318). Knowledge and understanding gained from materials science research, then, might generate new economic opportunities.

Another important way in which space research is likely to contribute to economic growth is the already mentioned idea that a requirement of various forms of space research is improved instrumentation. As Ralf-Jürgen Dettmar has argued,

> [p]rogress in observational astrophysics is—like in other empirical sciences— very closely linked to progress in measurement techniques. More specific, it is increased sensitivity and resolution with regard to the incoming (mainly) electromagnetic radiation that allows for the detection of new phenomena and for the better measurements which change our perception (2006, 174)

This point is reinforced by Hansjörg Dittus, who reminds us that in fundamental physics especially, "experimental accuracy is limited by space technology mainly" (2006, 294). He elaborates further:

> Beside tests of special and General Relativity, the investigation of the long list of open questions in cosmology, such as the search for dark matter and dark

---

19. See Gerzer, Hemmersbach, and Horneck (2006) for further information about life sciences research in space.

energy as well as the investigation of eventual anomalies of the Newtonian potential on long and short distances gives rise to plans for future missions and single experiments which can be mounted on any deep space probe. Challenging experiments set the goals, and the development of completely new technology for these experiments and missions is useful and promising. (294)

The pursuit of many forms of scientific research call for the development of new, improved, and increasingly sensitive instrumentation. This interplay between science goals and technology is an especially fruitful avenue for the generation of scientific progress in Douglas's sense, i.e., of the expansion of our powers of observation, prediction, and control.

Research That Contributes to Democratic Culture

The final category Brown and Guston identify captures many, if not all forms of space research, *viz.*, research that contributes to democratic culture by satisfying intellectual curiosity and offering aesthetic pleasure. Though I will focus here on items of scientific curiosity in space, it is worth noting the contributions of astronomy and Earth observation to the experience of aesthetic pleasure. Optical imagery has provided stunning images of our home planet, the other planets and moons of our solar system, as well as nebulae and galaxies.[20] Though the comparative value of these pictorial contributions may be difficult to ascertain, it is clearly non-negligible, as evidenced by the widespread public opposition to past attempts to end the operation of the Hubble Space Telescope, which captured, among many other notable images, the "Pillars of Creation" image from the Eagle Nebula (Fig. 3.1). We should also recognize the value of space science to many other art forms. Our increased awareness and understanding of our universe have vastly expanded on the possibilities for enjoyable and insightful novels, films, television shows, performances, music, etc.

There is an exceptionally broad array of scientific disciplines and topics of interest that can be addressed via space research, and in many cases, *only* via space exploration. Earth orbit is likely to remain an attractive location for astronomical observation, and not only because it is nearby. This is because Earth's atmosphere is opaque to much of the electromagnetic spectrum, and so much data of interest to astronomers and other researchers can only be acquired with space-based equipment, whether in Earth orbit or elsewhere

---

20. This work has also fueled artistic creation, with some of this artwork in turn inspiring space research. See Newell (2014) for an account of the origin and influence of Chesley Bonestell, who created many well-known space artworks, including *Saturn as Seen from Titan*.

Figure 3.1 The "Pillars of Creation" image of a section of the Eagle Nebula. Composite image from the Hubble Space Telescope.
Credit Line: NASA, ESA and the Hubble Heritage Team (STScI/AURA).

in the Solar System (such as the L2 point of the Earth-Sun system where the James Webb Space Telescope will be stationed). Low-Earth Orbit (LEO) is also an attractive setting for the study of Earth's magnetic fields. Meanwhile, highly eccentric Earth orbits provide the opportunity to engage in high-precision physics experiments involving changes in the gravitation potential (Dittus 2006).

Beyond Earth, our Solar System provides a multitude of environments, the study of which will help to answer many questions of contemporary interest. Examination of virtually every body in the Solar System will expand our understanding of the history and evolution of the Solar System. Of particular interest here is lunar exploration. Though the Moon is the second-most explored body in the Solar System, it is somewhat uncommon as an example of a body that has been geologically inactive for much of its existence. Examination of its structure and composition would provide extensive information about the details of the Solar System at the time of its formation. Moreover, since the Moon lacks any endemic surface recycling mechanisms, studying its extensive surface cratering can teach us about the bombardment history of the inner solar system, and provide additional geological context for existing lunar samples and meteorite fragments (Crawford et al. 2012, 4). Similar insights would become available through, for instance, the exploration of Phobos, the largest of Mars's moons (Murchie, Britt, and Pieters 2014). Perhaps surprisingly to some, examinations of the gas giants (Jupiter and Saturn) and the ice giants (Uranus and Neptune) are also likely to inform our understanding of the formation and evolution of the Solar System, including the inner Solar System. According to Mousis et al.,

> Giant planets contain most of the mass and the angular momentum of our planetary system and must have played a significant role in shaping its large scale architecture and evolution, including that of the smaller, inner worlds. Furthermore, the formation of the giant planets affected the timing and efficiency of volatile delivery to the Earth and other terrestrial planets. Therefore, understanding gas giant planet formation is essential for understanding the origin and evolution of the Earth and other potentially habitable environments throughout our solar system . . . . Comparative planetology of the four giants in the solar system is therefore essential to reveal the potential formational, migrational, and evolutionary processes at work during the early evolution of the early solar nebula. (2014, 30)

With respect to Saturn, they note that it would be especially informative to acquire data on: the planet's atmospheric temperature and pressure profiles; the abundances of carbon, nitrogen, sulfur, oxygen, helium, neon, xenon, krypton, and argon in its atmosphere; as well as the isotopic ratios of its various atmospheric gases such as hydrogen and helium (45). There is a similar value to the study of the ice giants, especially of their satellites, about which comparatively little is known (Turrini et al. 2014).

Geology and planetary science are likely to be advanced through the study of geological processes on bodies throughout the Solar System. Seismology experiments, which are important for understanding the structure of planetary interiors, have so far been conducted only on Earth and the Moon. The comparative study of various physical processes on other planets is also of interest, as Tobie et al. describe in the case of Saturn's moon Titan. Titan, which is the second largest moon in the Solar System (after Jupiter's moon Ganymede), is the only planetary satellite with an atmosphere capable of sustaining clouds. Since Titan is in many ways similar to Earth, it "offers the possibility to study physical processes analogous to those shaping the

Earth's landscape, where methane takes on water's role, and to analyse complex chemical processes that may have prebiotic implications" (Tobie et al. 2014, 60). Another of Saturn's moons, Enceladus, was discovered by the *Cassini* spacecraft to have plumes of water vapor jutting from its southern pole. The presence of water on Enceladus "provides a unique opportunity to analyse materials coming from its water-rich interior, potentially containing compounds of prebiotic interest, and to study today aqueous processes that may have been important on many other icy worlds in the past" (60). Thus, both Titan and Enceladus are of interest in the search for life (even if they are not themselves home to life), because their examination will teach us more about the nature and scope of habitability.

Regarding the search for life, Mars and Europa are also of interest. Mars was very likely habitable long ago in the past, and might contain traces of past life (or even enclaves of life in subsurface environments where liquid water exists). Europa, meanwhile, has a large subsurface liquid water ocean. A discovery of a second origin of life would constitute one of the most important discoveries in the history of biology, and might lead us to rethink various aspects of biology at a fundamental level. Even if these bodies are devoid of life, that does not diminish entirely their scientific significance, as Cockell argues in the case of Mars exploration:

> [I]f all of the most plausible candidates for habitats . . . show no evidence of life, then what biological exploration is there [to do]? . . . [M]any important biological questions would be invoked. For example, are there environments where the conditions for life are met (pH, water availability, redox couples, etc.), but where there is no life, i.e., habitable, but sterile environments? Were these environments, if they exist, ever habitable? What was different about early Mars compared to early Earth that precluded the origin or transfer of life to Mars, when we know that early Earth was conducive to life? . . . [I]f we took life to Mars and implanted it would it survive and grow . . . ? (2004, 234–235)

Analogues of these questions would gain importance should the exploration of Europa, Enceladus, and Titan reveal no signs of life. Cockell's last question intimates that there is scientific interest in questions about the survivability of terrestrial life in space. As he argues elsewhere, astrobiology stands to benefit from, e.g., studying the unsterilized remains of crashed or abandoned spacecraft on the Moon to see if any microbial life has survived (which would help determine the effectiveness of planetary protection protocols), as well as conducting long-term reproduction experiments on the Moon to identify potential effects of reduced gravity on organisms over multiple generations (Cockell 2010).

Also of relevance to the study of habitability and the search for life is the growing field of exoplanet detection. The first confirmed discovery of a planet orbiting another star occurred in 1992 (though unconfirmed detections date back to 1988). At the time of writing, a total of 4,075 exoplanets have been discovered, orbiting 3,043 stars, 660 of which have been confirmed to have multiple planets.[21] Such discoveries have been made possible by improvements in various detection methods. The goal of discovering smaller, Earth-like planets (and ideally the spectral and other analyses of their atmospheres) is of great interest and will spur future innovations in instrumentation. Other items of astronomical interest include the use of X-ray satellites to study matter under strong gravity, as well as the exploitation of wavelengths either not accessible from the surface of the Earth, or too "polluted" for useful study. A potentially valuable destination for optical and radio astronomy is the lunar farside. The Moon, which is seismically stable, allows for the placement of highly sensitive optical sensing equipment. Moreover, the farside is shielded from Earth's radio emissions. According to Paul Spudis, this would permit "observations in totally new parts of the radio spectrum" (2001, 160). And, Spudis claims, "[e]very time a new portion of the spectrum has been observed, new discoveries have been made" (160).

Admittedly, such a brief overview of space studies is woefully insufficient for articulating the scope and extent of scientific interest in the space environment. There is simply too much and too detailed of information for a philosopher to do justice to. However, I hope I have provided enough of a picture of the scientific potential of solar system exploration to support the main contention of this section—that space science has much to offer to our culture in the satisfaction of curiosity, but also in the provision of new ideas and new perspectives. Thus, it should be clear that the state has *prima facie* obligations to support space research of many kinds. If Brown and Guston are right about the relative strength of the state's obligation to support research, then the strongest obligations associated with space research attach to those projects that promise to help resolve political questions, with correspondingly weaker obligations associated with space research falling into the other categories. I do not straightaway accept this conclusion, if only because, as Matthew Brown has suggested, the *significance* of a research project is not a function merely of the *importance* of a practice but also the degree to which that practice has been disturbed (for instance, by anomalous observations). Thus, even for research that is not immediately politically relevant, its having been disturbed in some important way can increase its priority in comparison to other research projects. That various space sciences pose significant, unanswered

21. Exoplanet Catalog, http://exoplanet.eu/catalog/ (accessed 8 June 2019).

questions, adds to their importance, and thus to the degree to which the state is obligated to support research that would help to answer those questions.

Humans or Robots?

Heretofore I have attempted to remain agnostic regarding one of the more thorny disputes among space scientists, *viz.*, the wisdom of sending *human* researchers into space, as opposed to engaging in exclusively robotic forms of exploration. This debate has a noticeable disciplinary bias, with astronomers and astrophysicists tending to oppose human exploration; and with geologists and biologists tending to favor human exploration (at least when planetary protection considerations are moot). This disciplinary bias exists in part because the benefits of human exploration tend not to be apparent for astronomy and astrophysics, for which remotely controlled telescopes (and other equipment) usually suffice. Meanwhile, the benefits of human exploration are much more apparent when it comes to the collection of samples from planetary surfaces, where spur-of-the-moment insights are often more scientifically rewarding than pre-programed rover instructions. Given the degree to which scientists must specialize in the contemporary research environment, it is understandable why an astronomer and a geologist might come to blows over this issue.

At the risk of annoying all parties to the dispute, I intend to remain agnostic about the urgency of sending humans into space, at least for the time being. I agree with critics of human exploration that sending humans into space increases costs and risks in many unacceptable and downright unaffordable ways. It seems unlikely that, at present, a human expedition to a destination other than the Moon, a near-Earth asteroid, or Mars would produce scientific returns large enough to justify its great cost and its health risks to astronauts. Although their cost has increased to match their increase in capabilities, robotic missions remain cheaper for the time being, and can more easily explore environments lethal to humans. At the same time, in those situations in which humans have been present, the scientific returns have been extensive. The Apollo program, which for political purposes sent twelve humans (including one professional geologist) to the lunar surface, returned a total of 382 kg of samples. Meanwhile, the successful Soviet robotic sample return missions (Lunas 16, 20, and 24) altogether returned only 321 g of samples. The payoff of Apollo in terms of publication is also considerable. According to Ian Crawford, nearly 3,000 refereed publications can be traced to the study of the Apollo lunar samples (2012, 2.24). This number comes from a search of the Astrophysics Database System. Meanwhile, using similar parameters, a search of the GeoRef database, which focuses on the geosciences, revealed over 8,000 journal articles and books linked to the Apollo missions.[22] This supports Crawford's contention that "any space mission which has to transport people will, by its very nature, also be able to carry a significant scientific payload, even if science is not the primary driving force behind it" (2001, 156).

As Spudis explains in the case of lunar geological exploration, it is important to be mindful of different and complementary approaches to studying an environment. On the one hand exploration calls for *reconnaissance*, which provides a rough account of what exists in a given environment. Reconnaissance is largely observational and can in many cases be conducted effectively via orbital observation, landers, and rovers (as well as human explorers and teleoperators). Ideally reconnaissance gives way to *field study*, where "the object is to understand planetary or biological processes and history at whatever levels of detail are appropriate" (163). Effective field study is currently beyond the capabilities of robotic explorers:

> [Field study] requires observation in the field, the mental building of a conceptual model, hypothesis formulation and testing, done with repeated visits to the same geographic location. Field study is an open ended, ongoing activity; some field sites on the Earth have been studied continuously for over a hundred years and still continue to yield new and important insights. Field study requires the guiding presence of human intelligence and is not a simple matter of collecting data, but is the process of analyzing data in the field and applying these insights on a continuing basis to formulate increasingly more sophisticated and complex questions . . . .
>
> Field study is complicated, interpretive, and protracted. Moreover, the plan of problem solution is not immediately apparent before, and sometimes during, the field work but must be formulated, applied, and significantly modified in real time. Most importantly, field work nearly always involves uncovering the unexpected and in this type of work, discoveries can be exploited to a degree not possible during simple reconnaissance. Sometimes such exploitation calls for exploration methods and techniques completely different from those originally envisioned. (163–164)

In the absence of robotic explorers that rival the speed, dexterity, and interpretive insights of humans, field study of space environments will require human explorers (or at least, human teleoperators (Lester and Thronson 2011) stationed close enough to reduce communications delays to acceptable levels). Moreover, since field study is likely necessary for understanding other planetary environments to the degrees needed to support meaningful

22. I conducted this search on 15 June 2018, searching for "Apollo AND Moon" in titles, abstracts, or subject listings.

comparative planetology, it seems prudent that we remain open to missions with human explorers. That said, it does not follow that we are at the point of diminishing reconnaissance returns for any particular Solar System body. Perhaps we have engaged in sufficient reconnaissance of the Moon so that we can intelligently and effectively engage in lunar field study. However, the rest of the Solar System is probably insufficiently reconnoitered; there is still much of value that we can learn through continued reconnaissance on and around Mars and elsewhere.

We should also be mindful of direct, biological reasons for studying humans in space. I have already mentioned several questions of interest, mostly relating to the effects of reduced- and micro-gravity on organisms (including humans). According to Mark Shelhamer, such inquiries are not obviously any less important than "knowing the composition of galaxies" (2017, 38). One example of observations that simply would not have been possible without human explorers (as subjects) were the observed effects of long-duration spaceflight on human vision:

> This problem appears to arise from the head-ward redistribution of body fluids in space due to the lack of a net gravitoinertial force vector, which in turn produces a chronic increase in intracranial pressure, causing cerebrospinal fluid to impinge on the back of the eye and change its shape. *We would not know about this possible effect of prolonged fluid shift if the vision problems had not been reported by astronauts themselves.* (38–39; emphasis added)

Shelhamer's point is that this finding gives rise to further questions of legitimate scientific interest, not just about human physiology, but also about life in space more generally:

> What does this tell us about life that might exist elsewhere, in either reduced or increased gravity levels relative to Earth? What evolutionary selections might have occurred in different planetary settings to avoid this issue, which might have not only an effect on vision but possibly on neural function if maintained for longer durations? If we wish to consider not only the possibility of life beyond Earth but what forms it might take, here is an example of how human space flight can contribute to that discussion. (39)

Therefore, we should not immediately dismiss the value of sending humans into space, either to work as explorers, or to be studied as human subjects. The latter use of human explorers broaches a sundry of questions related to human subject research. Given the large number of unknowns, especially with respect to long-duration spaceflight, it is an open question precisely how best to characterize voluntary, informed consent to participate in long-duration human subject experiments in space. At the same time, it is unclear how to remove the risks and uncertainties without increased experimentation on humans during long-duration missions. On this matter I have little to add.[23]

The upshot is that, while we ought to appreciate the effectiveness and ingenuity of human explorers, it is not clear that there is presently a great scientific need to send humans to any new destinations in space. For the time being, increased robotic "reconnaissance" missions may yield the best science returns per science dollar spent. That does not mean we should not continue to make such progress as we are able to make in the study of human physiology in the space environment, for instance through Earth orbiting missions. It is simply to claim that there is no harm in forestalling human exploration to various destinations until we can be sufficiently confident both that reconnaissance of each destination has reached the point of diminishing returns, and that minimal harm will come to those humans participating in the missions.

ETHICS IN SPACE?

Up to this point in the book my primary concern has been to provide a positive rationale *for* space exploration, e.g., that we have an obligation to support various space research projects because they will provide us with understandings that are both intrinsically and instrumentally valuable. And while I hope I have provided compelling arguments, I grant that I have said little which might help us to determine the comparative strength of these obligations. So, it may well be that someone could agree with virtually everything I have said so far and consistently maintain that, for instance, there is no need to increase funding for space research. However, even if that is the reaction of the majority of readers of this book (though I hope it is not!), there is another, perhaps even more important role for the intrinsic and instrumental values of scientific understanding. And this would be their contributions to resolving ethical questions surrounding activities *within* the space environment. There are many non-scientific uses of the space environment, such as space resource exploitation, space tourism, and human settlement. Each of these activities has the potential to interfere with the scientific uses of space. For instance, a permanent human settlement on the Moon could substantially increase the levels of dust in its tenuous atmosphere, diminishing its value as a location for sensitive optical equipment. Already we face questions about whether and to what extent such activities should be permitted or encouraged. What

---

23. But see, e.g., Koepsell (2017) for a discussion of human subject concerns as they apply to the *Mars One* project, as well as my discussion of space settlement in Chapter 6.

I have said in this and the previous chapters has primarily been to establish science as a key stakeholder in the use of the space environment. As I shall argue over the course of the remining chapters of this book, other uses of the space environment are subject to comparatively weaker ethical justifications. Thus, for the foreseeable future, questions about the permissible use of space environments must be resolved in ways that maximally preserve opportunities for scientific exploration and study.

CHAPTER 4

The Scope and Justification of Planetary Protection

One of the most widely discussed aspects of space science policy concerns the scope, rationale, and requirements for the protection of sites of interest in the search for evidence of extraterrestrial life in the solar system. Dating back to the build up to the International Geophysical Year (1957–1958) (which saw the launch of *Sputnik 1*, the first artificial terrestrial satellite), concerns were raised about the possibility of biological contamination resulting from space exploration activities. Two basic challenges were recognized: On the one hand, there is a need to protect Earth's environments and their inhabitants from contamination from the space environment (a concern dramatized in Michael Crichton's *The Andromeda Strain*). This issue is known as *back contamination*. On the other hand, there is a need to preserve the viability of the search for evidence of life in extraterrestrial environments. Biological and organic chemical contamination of the space environment could result in false positive life signs, or at least in unnecessary ambiguities in any potential findings. This issue is known as *forward contamination*. Reflective of concerns about both back and forward contamination, the Committee on Space Research (COSPAR), established in 1958 by what was then known as the International Council of Scientific Unions, first developed a set of *planetary protection policies* in 1964. These policies were the inspiration for the language of Article IX of the United Nations Treaty on Principles Governing the Activities of States in the Exploration and Use of Outer Space, including the Moon and Other Celestial Bodies (Outer Space Treaty), which states in part that:

> States Parties to the Treaty shall pursue studies of outer space, including the Moon and other celestial bodies, and conduct exploration of them so as to avoid

their harmful contamination and also adverse changes in the environment of the Earth resulting from the introduction of extraterrestrial matter and, where necessary, shall adopt appropriate measures for this purpose.

The Outer Space Treaty forms the foundation of binding international space law; however, it is notable that the document nowhere provides a definition of "harmful contamination." The specific requirements of planetary protection policies, though proffering an account of "harmful contamination," do not carry the force of law, even though they have been implemented by space programs engaging in interplanetary exploration. COSPAR has actively updated its policies as new information has become available about which environments are likely candidates in the search for evidence of life, and also as new information has become available about spaceflight capabilities and contamination risks.

Although I will provide an overview of the current COSPAR requirements in Planetary Protection Policies, my focus in this chapter will be on the *rationale for* and *scope of* planetary protection against *forward* contamination. It has been widely recognized that, should any extraterrestrial life exist, the potential impact to that life of our exploratory activities would raise distinctively ethical questions about the value of, as well as our duties toward, extraterrestrial lifeforms. It has consequently been proposed that planetary protection policies have an ethical dimension, including an obligation to protect extraterrestrial life for its own sake, even if that life is microbial. This has led a number of astrobiologists and philosophers to search after an account of the intrinsic value of life—one that is capable of validating claims about, e.g., the intrinsic value of any microbial life we might discover on Mars, Europa, or Enceladus. Charles Cockell has put forward the most thoroughly developed account of microbial value, which I discuss in Ethics for Life. Cockell's position is, in its most recent articulation, a kind of non-anthropocentric, holistic environmental ethic that generates obligations to microbial communities or ecosystems.[1] Though I am ambivalent about the particular way in which Cockell attributes intrinsic value to microbial life, I grant that his defense of the value of microbial life is compatible with the approach to attributing intrinsic value I developed in Chapter 2.

---

1. An environmental ethic is *anthropocentric* when it regards humans as the only (or as the primary) loci of intrinsic value (or as the only things worthy of moral consideration), implying that much or all of the nonhuman world is at best instrumentally valuable. An environmental ethic is *non-anthropocentric* when it recognizes at least some nonhuman entities as intrinsically valuable (or morally considerable). Such an ethic is *holistic* if it grants value or moral consideration to collective entities (such as species or ecosystems); and *individualistic* if it grants value or moral considerations only to individuals.

Cockell's microbial ethics approach to planetary protection points to a common, but problematic feature of deliberation about planetary protection. There is a tendency to view protection against forward contamination as motivated *ethically* only when it is based upon a respect for the value of extraterrestrial life. Such an attitude has the implication that there can be nothing genuinely *ethical* about the protection of space environments that are not home to life. In Ethics for Non-Life I explore various attempts to reject this attitude, which I have elsewhere (Schwartz 2019b) termed the "life bias." I consider two strategies. First is Holmes Rolston III's account of intrinsic value, which applies to space environments regardless of whether they are home to life. Second is Tony Milligan's view that it is constitutive of a good human life that we value the integrity of space environments, whether home to life or otherwise. Although I approach each of these positions with caution, I do not intend to refute them. Instead, I describe them because they articulate a commendable attitude toward the space environment, *viz.*, one that does not denigrate space environments simply because they lack certain valued features of terrestrial environments. Still, each view is guilty of some degree of underestimation of the potential for scientific values to justify broad protection of the space environment. It should be evident to readers of the previous two chapters why we have an ethical obligation to engage in the scientific examination of the space environment. As an extension of this, I argue that we have a corresponding obligation to protect sites of scientific interest, which includes not only sites of interest to astrobiologists, but also sites of interest to those of many other disciplinary orientations. Since space science has yet to disclaim interest in any particular space environment, the requirements here are broad and powerful.

I close the chapter in Interstellar Implications with a brief discussion of long-term planetary protection issues. Given the rapid progress of the Breakthrough Starshot initiative, which might realize interstellar exploration using gram-scale probes on the order of *decades*,[2] it is wise to initiate discussion about planetary protection for interstellar travel. I argue that planetary protection policies should be extended to include interstellar missions, and I offer some very provisional recommendations for the protection of scientific interests on these missions.

PLANETARY PROTECTION POLICIES

In its most recent formulation, COSPAR's planetary protection policies specify protocols for missions falling into five different categories. Category

---

2. See Lubin (2016) for an overview of the technologies involved and their readiness levels.

I missions are those "to a target body which is not of direct interest for understanding the process of chemical evolution or the origin of life" (Kminek et al. 2017, 13–14). At present these include flyby, orbiter, and lander missions to Io (the innermost of Jupiter's four Galilean moons) and to most asteroids. No protection requirements are imposed because "[n]o protection of such bodies is warranted" (14). Next are Category II missions, which "comprise all types of missions to those target bodies where there is significant interest relative to the process of chemical evolution and the origin of life, but where there is only a remote chance that contamination carried by a spacecraft could compromise future investigations" (14). These include flyby, orbiter, and lander missions to Venus, Jupiter, Saturn, Uranus, the Moon, Ganymede and Callisto (moons of Jupiter), Titan (Saturn's largest moon), Pluto and its largest moon Charon, Ceres, and Kuiper-Belt objects. The only requirement for such missions is "simple documentation" of potential impact locations. Category III missions are those "to a target body of chemical evolution and/or origin of life interest and for which scientific opinion provides a significant chance of contamination which could compromise future investigations" (14). These include flyby and orbiter missions to Mars, Europa (a moon of Jupiter), and Enceladus (a moon of Saturn). Requirements include documentation over and above that required for Category II missions, in addition to the possible use of trajectory biasing, cleanroom assembly of flight hardware, and bioburden reduction.

Category IV missions include those "to a target body of chemical evolution and/or origin of life interest and for which scientific opinion provides a significant chance of contamination which could compromise future investigations" (14), and these include lander missions to Mars, Europa, and Enceladus. Requirements here are more extensive and involve:

- A bioassay to measure the bioburden of flight hardware.
- A probability of contamination analysis.
- An inventory of constituent organics.
- Additional procedures such as trajectory biasing, cleanroom assembly, bioburden reduction, and partial sterilization of lander hardware.

More specific Category IV requirements have been specified for Mars and for Europa and Enceladus. These vary based on the purpose and intended destinations for missions to these bodies. For instance, on Mars there are areas specified as "special regions," which are areas in which either "terrestrial organisms are likely to replicate" or that are thought to have a "high potential for the existence of extant martian life" (18). Areas currently identified as special regions are those containing gullies, subsurface cavities, the subsurface below 5 m, recurrent slope lineae, as well as any locations found to have groundwater, methane sources, or geothermal activity. Missions to these regions require much more extensive bioburden reduction of the entire landed system, in addition to sterilization of all contact hardware.

Lastly, Category V missions include all Earth-return missions, in which the concern is to protect the Earth and the Moon from back contamination.[3] These missions are divided into two further groups—restricted Earth return missions, and unrestricted Earth return missions. Unrestricted Earth return missions include those to Venus and the Moon, and involve protection requirements only for the outbound phase. Restricted Earth return missions include those to Mars and Europa. In addition to forward protection requirements, these missions also require "the absolute prohibition of destructive impact upon return, the need for containment throughout the return phase of all returned hardware which directly contacted the target body or unsterilized material from the body, and the need for containment of any unsterilized sample collected and returned to Earth" (14–15).

My focus will be exclusively on the scope of and rationale for protection against *forward* contamination. Thus, any subsequent use of the term "planetary protection" should be understood to mean "planetary protection against *forward* contamination." In restricting my focus to the issue of forward contamination I do not mean to suggest that protection against back contamination is trivial or insignificant, or that it is of marginal ethical moment. It is only that I see little controversy, as far as its ethical justification is concerned. No matter what humans do elsewhere in the Solar System, there remains an obligation to protect the terrestrial ecosystems on which human and other life on Earth currently depends. Consequently, it is clear why we ought to protect Earth against contamination from extraterrestrial material. To the extent there is any controversy here at all, it has mostly to do with the level of caution appropriate for handling samples possibly containing extraterrestrial organisms. What are the odds, if introduced to terrestrial ecosystems, that these organisms would become a destabilizing influence? Prudence counsels caution, given how much is at stake.

What of the rationale for COSPAR's forward contamination protection policies, then? According to the preamble of COSPAR's planetary protection policy, protection is needed because "[t]he conduct of scientific investigations of possible extraterrestrial life forms, precursors, and remnants must not be jeopardized" (13). Historically this motivation has been viewed as a purely scientific consideration. This attitude is hard to defend, if only because it is difficult to maintain that considerations about the proper conduct of science are outside of the scope of ethics—or that ethical judgments are outside of the scope of science. Indeed, value judgments *pervade* science, not only in decisions about *how* to carry out investigations, but also in decisions about

---

3. The reason for including the Moon is to "retain freedom from planetary protection requirements on Earth-Moon travel" (14).

what is worth investigating in the first place. It is also the case that the results of scientific investigations affect value judgments because these investigations often offer us new insights into the nature of phenomena. These new insights sometimes cause us to reevaluate our beliefs about what exists and about what factors we take to be significant when engaging in ethical deliberation, but also in the everyday task of living our lives. There is, then, a strong mutualism between scientific considerations and considerations of ethical valuation and theorizing.[4] For this reason, decisions about the conduct of scientific exploration cannot be dismissed as "purely scientific" or "purely pragmatic," even if those engaged in scientific deliberation do not consider themselves to be engaged in projects that include ethical components.

But even if there is ample room within the scientific process to recognize a role for ethical deliberation, that does not guarantee that scientific interests are exhaustive when it comes to the ethics of planetary protection. Two developments are important here. First, research in the 1980s and 1990s into the feasibility of terraforming (in particular, the terraforming of Mars) quickly led certain researchers to consider whether terraforming, if feasible, *ought* to be done. According to Christopher McKay (1990), prior to instigating terraforming we must first resolve the question of whether Mars has indigenous life. In particular, McKay held that the existence of indigenous life either should *preclude* terraforming (if it would harm that life), or should *encourage* terraforming for the purpose of helping that life thrive (rather than for the purpose of providing humans with a new home planet). A basis of McKay's perspective is that Martian life, if it exists, would be valuable for its own sake and thus would be morally considerable. Second, in 1996 researchers presented possible evidence of indigenous Martian life from a study of the Alan Hills 84001 meteorite. A key finding here was the presence of magnetite formations that strongly resemble biogenic magnetite, but that can be produced by non-biological means (albeit with greater difficulty). Though the astrobiology community is still divided over the strength of the evidence in this case, the perceived likelihood of discovery of Martian life increased, and the following years saw a marked increase in the amount of discussion about the moral considerability of extraterrestrial life. This led to growing acceptance of the idea that planetary protection policies must be implemented for the protection of extraterrestrial life *for its own sake*.

The upshot is that it is quite common in the astrobiology community—at least, common among those who discuss rationales for planetary protection—to find acceptance of the intrinsic value of extraterrestrial life, even if that life is microbial in nature. But from a normative ethical perspective such attributions of intrinsic value are contentious. After all, one cannot *merely*

---

4. See Schwartz (2016a) for a substantive defense of this claim.

assert that such life is intrinsically valuable. Ideally, such value claims should be consequences of an attractive and coherent background normative ethical theory, one that informs decisions broadly speaking and not just those about planetary protection. The problem, however, is that no traditional ethical theory (e.g., Kant's deontology; Mill's utilitarianism; Aristotle's virtue ethics) implies that microbial life is intrinsically valuable. Even within environmental ethics there is exceptionally little discussion of microbial intrinsic value. (Although it would be easy for an ethicist of any stripe to recognize the *instrumental* value of microbial life.) What, then, should be said *in defense* of the idea that microbial life, including extraterrestrial microbial life, is intrinsically valuable, and that *prima facie*, it is worth protecting for its own sake?

ETHICS FOR LIFE

The question I would like to consider first, then, is what justification should be given for thinking that extraterrestrial microbial life is intrinsically valuable. A reasonable starting point would be to adopt some relatively permissive or wide-ranging environmental ethic, such as Aldo Leopold's (1949) Land Ethic. The Land Ethic is a non-anthropocentric, holistic environmental ethic that locates the source of intrinsic value at the level of ecosystems or, to use Leopold's terminology, "biotic communities." His famous maxim is that "[a] thing is right when it tends to preserve the integrity, stability, and beauty of the biotic community. It is wrong when it tends otherwise" (224–225). Extraterrestrial microbes, if they exist, would most likely belong to their own biotic communities, and thus, our actions could be evaluated on the basis of their impact on (or respect for) the integrity, stability, and beauty of these communities.[5]

However, if one contemporary champion of the Land Ethic, J. Baird Callicott, is correct, then extraterrestrial ecosystems fall outside the scope of the Land Ethic. On Callicott's interpretation, the Land Ethic derives its normative force from humanity's rough biological kinship with the rest of life on Earth. Not only do we share an evolutionary heritage with most or all terrestrial life forms, but we also are each members of the same terrestrial biotic community. Thus, it is appropriate, if not evolutionarily expected, that we possess something like a Humean natural sympathy or empathy for these

---

5. Or, to avoid a kind of terrestrial-chauvinism, we might say that our actions would be evaluated on their impact on (or respect for) whatever factors are essential to the thriving of these communities, since concepts like "integrity," "stability," and "beauty" (at least as we ordinarily understand them) may be unsuitable in extraterrestrial contexts.

other members of our community. This poses a problem for the consideration of extraterrestrial life, because extraterrestrial life forms,

> assuming they were not of Earthly origin and inoculated somehow on some foreign body, or *vice versa*, would not be our kin—that is, descendants of a common paleontological parent stock—nor would they be participants in Earth's economy of nature or biotic community. Hence, they would lie outside the scope of Leopold's land ethic. (Callicott 1986, 246)

Callicott is careful to explain that the terrestrial scope of the Land Ethic is not a disproof of the intrinsic value of extraterrestrial life; rather, his point is that the Land Ethic "simply has nothing to say about extraterrestrial life" (247). Still, Callicott's exclusionary move is unnecessary. To begin with, we might cast doubt on the idea that kin-relations, even distributed over evolutionary time, are necessary for sympathizing or empathizing with biotic communities. That it is within the capacity of our moral sensibilities to feel this way about extraterrestrial communities can be seen in a number of ways, and most starkly by our sympathetic reactions to the plight of alien species in various science fiction stories, even in stories set in universes with no humans or where humans are not members of the depicted communities of life.[6] While this might be more difficult to maintain in the case of microbial extraterrestrial communities on Mars or elsewhere in the Solar System, it is not impossible or incoherent that humans could sympathize with these communities. We might even find an entirely novel basis for community identity, e.g., as members of the same solar system.

Alternatively, we might reject the idea that shared community membership is necessary for the recognition of value. Even if extraterrestrial microbial communities are evolutionarily disconnected from terrestrial biotic communities, it does not follow that we have no basis for valuing the former. For instance, these communities might share features in common with the terrestrial communities that we value. As Erik Persson appreciates, "[w]e have no problem finding reasons to value things—living and nonliving—as ends here on Earth, and we have no reason to believe that it will be any different with extraterrestrial life" (2012, 982). Moreover, we might find value in the ways in which extraterrestrial biotic communities differ from terrestrial biotic communities. We might, for instance, respect alien life for its ability to accomplish what we cannot—to survive unaided on Mars, Europa, or Enceladus. Little, then, prevents us from adopting extraterrestrial iterations of Leopold's maxim. Is there more to be said about why we should value extraterrestrial

---

6. This raises interesting questions about how we understand emotional reactions to fiction; see Yanal (1999).

microbial communities? Charles Cockell has made various proposals here that are well worth considering.

Cockell on Microbial Value

Cockell's account of microbial value belongs to the biocentric family of environmental ethical theories. What ties this family of views together is the identification of individual organisms as loci of intrinsic value. A useful example here is Paul Taylor's (1981) idea of *respect for nature*. Taylor's view, in outline, is that human persons are only one kind of member of Earth's wider community of life, a community in which all members are connected in complex and interdependent ways. For this reason, all individual organisms, human and nonhuman alike, are "teleological centers of life" and thereby are intrinsically valuable. It also follows, claims Taylor, that humans and human interests are in no way superior to nonhumans and their interests. In lieu of developing and assessing Taylor's overall position I would like to focus only on his method for connecting being a teleological center of life to possessing intrinsic value.

For Taylor, to view an organism as a teleological center of life involves recognizing that it has a "good of its own":

> To say that an entity has a good of its own is simply to say that, without reference to any *other* entity, it can be benefited or harmed. One can act in its overall interest or contrary to its overall interest, and environmental conditions can be good for it (advantageous to it) or bad for it (disadvantageous to it). What is good for an entity is what "does it good" in the sense of enhancing or preserving its life and well-being. What is bad for an entity is something that is detrimental to its life and well-being. (199)

While it is more or less clear what it means to speak of something as being good for a human person, it is less clear, at least on the surface, what it means when speaking about the "interests" of nonhumans, many of which lack the capacity to be *interested* in things. For this reason, Taylor is careful to explain that he does not refer only to conscious interests, but more generally, that we can think of the good

> of an individual nonhuman organism as consisting in the full development of its biological powers. Its good is realized to the extent that it is strong and healthy. It possesses whatever capacities it needs for successfully coping with its environment and so preserving its existence throughout the various stages of the normal life cycle of its species. (199)

Taylor then invokes his biologically permissive account of an individual's interests/good of its own in his account of intrinsic value:

> [R]egardless of what kind of entity it is in other respects, if it is a member of the Earth's community of life, the realization of its good is something *intrinsically* valuable. This means that its good is prima facie worthy of being preserved or promoted as an end in itself and for the sake of the entity whose good it is. (201).

Thus, each living organism, since it possesses a good of its own, it is thereby intrinsically valuable.

On structurally similar grounds Cockell has argued for the intrinsic value of microbial individuals (2005a). As he explains in a recent articulation of his position,

> [T]hat microbes have intrinsic value could be based on their possession of rudimentary interests. We know what is good or bad for a microbe based on physiological attributes . . . . [M]icrobes have latent tendencies and evolutionary capacities that might demand from us an appreciation of a value in them that transcends their use as resources. (2016a, 170)[7]

Thus, Cockell would insist that individual microbial organisms are loci of intrinsic value. However, he appreciates that this value claim would be difficult to apply in actual ethical deliberation, at least most of the time, because it is impossible for humans to exist without each day using and destroying trillions upon trillions of microbial organisms. For this reason, Cockell grants that a focus on microbial *communities* and their value (either as holistic entities or simply as the sums of their individual members) may be more appropriate. Such a respect, he claims, could be motivated by "the awe we feel for the sheer scale of [microbial communities'] biogeochemical processes and their longevity on Earth" (2016a, 171). But additionally, we might without apparent loss attribute interests to microbial communities. These communities as well as the larger ecosystems of which they are parts have tendencies and directions that could be altered or frustrated, providing the basis for considering what might be good or bad for these communities. If so, then we can identify microbial communities as loci of intrinsic value independent of any value we might place on the individual members of these communities.

---

7. *Cf.* Wilks (2016), who identifies "self-organization" as the value-conferring property that microbial life can be said to possess.

How should the recognition of the value of microbial life—either of individuals or communities—affect our conduct? Cockell proposes three guidelines:

1. We have a duty to preserve individual microbes except when doing so puts "constraints on human activities that are considered to be part of daily life" (172).
2. We have a duty to protect ecosystems and communities of microbial life (173).
3. We have "a general duty to show respect towards microscopic organisms in our activities" (173).

Cockell does not view the intrinsic value of microbial individuals or communities as overriding or as generating absolute, inviolable duties. Instead, he views these values and duties as *prima facie*, which might in certain situations be overridden quite easily. This has clear implications for planetary protection, however, because the destruction of microbial life on Mars, Europa, or Enceladus would certainly not be a part of any person's daily life, at least as things currently stand.

Should we adopt Cockell's position as an ethical basis for planetary protection policies? There are two sources of criticism of his position that should be considered. First, we might be skeptical about the principle on which those like Cockell and Taylor justify attributions of intrinsic value. And second, we might be skeptical about the use of intrinsic value in the first place.

Cockell and Taylor's positions each infer intrinsic value from the possession of interests, i.e., from the fact that living things (including microbial life forms) have goods of their own. The trouble with this is that it seems perfectly coherent for us to agree to the following two claims:

4. For each organism, something could be good or bad for it.
5. At least some organisms are not good in themselves.

Recognizing that individual life forms have interests is not tantamount to acknowledging that these life forms are intrinsically valuable. John O'Neill states this point nicely:

> It is possible to talk in an objective sense of what constitutes the goods of entities, without making any claims that these ought to be realised. We can know what is "good for X" and relatedly what constitutes "flourishing for X" and yet believe that X is the sort of thing that ought not to exist and hence that the flourishing of X is just the sort of thing we ought to inhibit . . . . That Y is a good of X does not entail that Y should be realised unless we have a prior reason for

believing that X is the sort of thing whose good ought to be promoted. (O'Neill 1992, 131–132)

In order to recognize the intrinsic value of microbes we must know more than merely that they possess rudimentary interests. We must have a reason for thinking that microbes are things "whose good ought to be promoted." Cockell's response to this kind of demand is to emphasize the scope and prowess of microbial life, which has "mastered and influenced the surface of the Earth in profound ways" (Cockell 2016a, 171). Whether this response is adequate depends on whether we should accept that we ought to grant intrinsic value to things which master and influence the surface of the Earth in profound ways. And even if we accept this, small enclaves of microbes in the Martian subsurface would not share with terrestrial microbes this kind of mastery. So, it is not clear that Cockell has highlighted what, if anything, would command our respect of *extraterrestrial* microbes, especially if their accomplishments are not noteworthy in comparison to those of terrestrial microbes. That is not to say no bases of respect could be found. Indeed, I have already proposed that we might respect extraterrestrial microbes for their ability to accomplish what humans so far cannot—to survive unaided in the hostile environments of space.

Although some might doubt that anything said above counts as evidence for the intrinsic value of microbial life, I suspect that many will to some extent share the intuition that microbial life forms, including extraterrestrial microbial life forms, manifest at least some laudable characteristics and so should be valued for what they are in themselves. Trouble remains, however, because one could agree with this and yet doubt that the intrinsic value of microbial life is in any way significant when it comes to resolving moral questions, including those pertaining to the protection and use of the space environment. Kelly Smith, who is broadly skeptical of intrinsic value claims as they apply to nonhumans, provides a voice for these concerns. Smith's objections are driven by a conception of intrinsic value on which the possession of intrinsic value is a "trump card against any attempt to argue that we should ignore the welfare of such an entity in favor of some instrumental use" (2009, 268). According to Smith, if two entities have intrinsic value then they are in a sense *equally* valuable, which significantly complicates ethical deliberation in hard cases, for instance, in lifeboat situations where we must make decisions about who will live and who must die. This poses a problem for Cockell, because if microbes are intrinsically valuable, then their value is on a par with the intrinsic value of humans. When faced with a decision between saving the life of a human and saving the life of a microbe, we would have no clear reason for preferring saving the human. But that is contrary to our assuredly justified intuitions about such a case, in which we would clearly favor saving the human over saving the microbe. For this reason, thinks Smith, we should reject the idea that microbes have intrinsic value.

It appears, however, that Smith and Cockell (as well as myself) are working with different conceptions of intrinsic value. For Smith, intrinsic values are, for all practical purposes, infinite; whereas for Cockell, intrinsic values come in degrees. Why does Smith adopt such a strong account of intrinsic value? He traces his own notion of intrinsic value to Immanuel Kant's (1993) conception of dignity. For Kant, all rational beings are equal with respect to their dignity, and thus, with respect to their intrinsic value. Smith explains the attractiveness of this kind of perspective:

> Traditionally, philosophers think of intrinsic value as a kind of line in the sand. Those things with intrinsic value are the fundamental units of ethical analysis. Instrumental value is viewed as either irrelevant or at least of decidedly secondary importance. One reason this is enormously important is that the values of items which have a merely instrumental value are subject to the forces of the market and thus fluctuate greatly depending on external circumstances. We certainly do not want the ethical value of humans to vary with the availability of plumbing or the whims of those who surround us at any particular time. So it's vital to forestall any instrumental calculation of human value by showing how it is completely beside the point from an ethical point of view—humans have instrumental value, to be sure, but that's not why we should treat them as morally important. As Immanuel Kant puts it, "People have a dignity, not a price." (2009, 266)

Without dissecting Kant's own reasoning, we can at least appreciate the need to see all humans as on a par, ethically speaking, and that the intrinsic value of human persons does not vary from person to person. Here we must recall that it is a feature of Kant's view that *only* rational beings are intrinsically valuable, and from experience the only *sufficiently* rational beings we know of are human persons. So of course, on Kant's view, all intrinsically valuable entities are as intrinsically valuable as human persons, since all and only human persons are intrinsically valuable.

But, as I have broached already,[8] Kant's is not the only conception of intrinsic value. Kant's conception of intrinsic value seems most useful for establishing certain kinds of absolute or overriding duties towards bearers of intrinsic value—for instance, that rational agents are to be treated always as ends and never as means. In other words, we have obligations towards "Kantian" intrinsically valuable entities because we have a duty to *respect* these entities for *what* they are in themselves. One could reject this picture

8. See Chapter 2, note 1.

entirely, or think that intrinsic values have other uses. For instance, on something like G. E. Moore's (1903) conception of intrinsic value, our duties regarding entities with intrinsic value are not duties to respect these entities as ends in themselves, but rather to protect and promote the good that they bring into the world. So, for someone like Moore, the point of intrinsic value is to mark off which entities make the world an overall better place simply for existing. This is, for better or ill, the kind of intrinsic value that I had in mind in Chapter 2 when arguing for the intrinsic value of epistemic states such as knowledge and understanding. Perhaps it is this Moorean kind of intrinsic value that Taylor and Cockell have in mind—it would not be surprising were Cockell to agree that, *ceteris paribus*, microbial lifeforms make the world an overall better place simply for existing.

Nevertheless, Cockell speaks as though we have duties *regarding* microbial life as an end in itself—that is, he thinks we have obligations *to* microbial lifeforms, e.g., that we have to respect their interests (and not just that we have to recognize the value they bring into the world). Thus, there is a Kantian-like element to Cockell's views about our obligations to microbes. This, then, represents a different kind of departure from Kant's conception of intrinsic value, which insists that the possession of intrinsic value (or a dignity) is not limited to rational agents but to anything that possesses rudimentary (or stronger) interests. Presumably, this means that there can be variation in the strength or complexity of an entity's interests, with this being correlated with the strength of our obligation to respect those interests. Such a view, it would seem, retains the attractiveness of Kantianism when it comes to resolving competing human interests, but furthermore brings other entities to the table of moral deliberation, and does so for their own sakes, in recognition of their own intrinsic values. So, it seems a rather simple riposte is available to Cockell: to insist that microbes are intrinsically valuable, just not to the same degree as human persons. Smith anticipates as much:

> The problem here is that such a move robs the intrinsic value account of its original motivation. The whole point of distinguishing intrinsic from instrumental value in the first place was to prevent any kind of haggling over who is more important than whom in different circumstances. In the eyes of a traditional defender of intrinsic moral value like Kant, to even enter into a discussion about which intrinsically valuable entities are more valuable than others is to have missed the whole point of moral value in a fundamental way. (270)

But this response is ineffectual precisely because it simply reaffirms the traditional Kantian perspective, and provides a motivation that Cockell, myself, and others need not share, *viz.*, that the "whole point . . . in the first place was to prevent any kind of haggling" about decisions involving intrinsically valuable entities. This motivation may be appropriate when *only* humans are being considered—but it should not be assumed in other contexts. As I put forward in Chapter 2, and as Cockell holds as well (2016a, 172),[9] intrinsic value is not a magical ethical wand that immediately resolves heretofore ambiguous cases or that fabricates overriding obligations of protection or preservation. Instead, Cockell can claim that an entity being intrinsically valuable merely provides this entity with a seat at the table of moral deliberation. All this establishes is that this entity must be taken into consideration for its own sake, even if it cannot speak for itself, and even if its value is vanishingly small. Although the outcome of deliberation may be unaffected sometimes (or even very often, for instance in cases where we must decide between humans and microbes), the nature of that deliberation is different. After all, being told that one is important, that one matters, just less so than others, is very different from being told that one is not important or that one does not matter at all (for one's own sake).

Smith also worries that Cockell's perspective is jaundiced by his disciplinary background (microbiology and astrobiology) and his focus not on ethics proper but on the narrower, and allegedly more controversial, subdiscipline of environmental ethics. According to Smith,

> [T]he environmental ethics movement has historically suffered from a kind of selection bias that affects its character in an odd way. The fact of the matter is that most people who specialize in environmental ethics have a deep interest in, even a reverence for, natural things. As a result, there's an understandable push to establish the significance of non-human "things" in the environment: animals, planets, ecosystems, mountains, rivers, etc. . . . . [T]he treatments of these questions in the environmental ethics textbooks that non-experts consult for authoritative opinions tend to have a definite point of view that is not necessarily representative of the ethics community more generally, which has for a very long time indeed tended to view that reason is morally special . . . . As a result, non-experts often take relatively radical ethical claims as uncontroversial. (2009, 267)[10]

The trouble with accusations of bias is that they are capable of cutting both ways. We could, with justification, lament the ecological ignorance of most ethicists as responsible for their tendency to discount the value of nonhumans. Theories of environmental value have not yet been given an adequate hearing within the profession of ethics, so it is unfair to denigrate these views as "radical." Moreover, most environmental ethicists are aware of and understand the

---

9. See also Randolph and McKay (2014).
10. Smith has gathered provisional evidence indicating opposition to a position like Cockell's among professional ethicists; see Smith (2016a).

more "traditional" theories and issues in ethics, whereas non-environmental ethicists are less likely to be aware of and understand environmental ethical theories. In this sense, environmental ethicists are probably more, rather than less, informed about ethics when compared with their non-environmental counterparts. It is also important to recognize the transformative character of knowledge, understanding, and specialization. As one studies a phenomenon, becomes more aware of it, and comes to understand it more and more, it is not unusual for there to emerge a sense of value for the phenomenon. But rather than greeting such value claims with skepticism, I would instead recommend that they be approached as learning experiences, that is, as opportunities to appreciate the significance of phenomena that we might not have the ability or time to develop on our own. That does not mean specialists are inerrant in their attributions of value and significance, and that we should blithely agree with their pronouncements about intrinsic value. But it does mean that, more often than not, such pronouncements are worth taking seriously and are worthy of critical investigation.

Whatever the ultimate merits of Cockell's position, or of other attempts to establish the intrinsic value of microbial life, we might nevertheless question whether more is at stake when it comes to planetary protection. One of Cockell's earlier musings might be taken as grist for the mills of those who deny any ethical motive for the protection of lifeless environments:

> "[L]iving" planets, even ones with unpleasant viruses, might be preferable to lifeless ones. If such an ethical position is adopted, we can argue that contaminating a dead planet lends value to that world . . . . We could, then, establish a reasonable ethical basis to conclude that failing to undertake any sterilization protocols for craft bound for lifeless planets is a *good* thing, as it increases the intrinsic value of the world being explored . . . . We might even deliberately *increase* the bioburden on a craft to help increase the value of other planets. Planetary protection might, in these rare instances, actually constitute a failure of our potential as moral agents to increase the intrinsic value of other planetary bodies, and we would end up in the somewhat bizarre position of concluding that planetary protection policies were unethical. (2005b, 291)[11]

Must the idea that life is more valuable than non-life lead to such conclusions? If we discover that Mars is lifeless, would that mean we ought to contaminate the planet with terrestrial organisms so that it becomes life-harboring? As I shall argue in the next section, these ideas should be rejected. Although we can grant

---

11. *Cf.* Cockell (2006). Cockell has clarified in personal communication that in describing the view "that planetary protection policies [are] unethical" as a "bizarre position" he intends neither to endorse this view nor to deny the possibility of non-life-oriented approaches to planetary protection.

that life contributes significantly to the value of a place, including its intrinsic value, we should stop short of saying that the presence of life is the *only* legitimate reason for protecting a space environment. There are good reasons for protecting space environments regardless of whether they are home to life.

ETHICS FOR NON-LIFE

It has not gone unrecognized that there is a need to consider reasons for protecting space environments regardless of whether they are home to life. Indeed, in 2010 COSPAR held a workshop on this issue, resulting in a series of recommendations related to the expansion of both the scope of and rationale for planetary protection policies. Two of these recommendations are especially salient here, including the first recommendation that an

> *expanded overall framework for COSPAR planetary protection policy/policies is needed* to address other forms of "harmful contamination" than currently addressed (*i.e.*, biological and organic constituent contamination). Such policy framework should be developed within the scope of the UN Outer Space Treaty (Article IX on harmful contamination). (Rummel, Race, and Horneck 2012, 1020; emphasis in the original)

The possible natures of these "other forms" of harmful contamination are specified in the fifth recommendation that the preamble of COSPAR's planetary protection policies be amended to acknowledge that

> *life, including extraterrestrial life, has special ethical status and deserves appropriate respect because it has both intrinsic and instrumental values* . . . . *non-living things, including extraterrestrial things, likewise have value and deserve respect appropriate to their instrumental, aesthetic, or other value to human or extraterrestrial life.* (1020; emphasis in the original)

These recommendations are noteworthy both for their recognition of the possible intrinsic value of extraterrestrial life, and for reserving a role for the consideration of non-living things. They do, however, raise questions about how to justify these value claims. I have already explored an account of the intrinsic value of extraterrestrial life; now I would like to consider what should be said about the value of "non-living things" in space.

Two basic approaches to the value of space environments, unrelated to the presence (or not) of extraterrestrial life, have been offered. First are proposals according to which space environments are themselves intrinsically valuable, and so ought to be respected, protected, and preserved for their own sake or for the good they contribute simply by existing. This is the strategy favored by Holmes Rolston (1986), Keekok Lee (1994), and Alan Marshall (1993). Second are character- or virtue-oriented proposals, according to which respect for space environments, even those not home to life, is one demand of living a good or virtuous life. This is the strategy favored by Robert Sparrow (1999; 2015), Saara Reiman (2009; 2010; 2011), and to a certain extent Tony Milligan (2015a; 2015b; 2018). Below I will focus on the views of Rolston and Milligan, respectively, since each provides the most developed examples of their respective strategy.[12] Though I shall not straightaway endorse or reject either perspective, I believe that each offers important insights and helps motivate and support my own favored rationale for planetary protection—the preservation of opportunities for the advancement of scientific understanding.

Rolston

Rolston is most well known as a founder of modern environmental ethics. Perhaps less well-known to environmental ethicists is his lasting impact on thinking about environmental issues in space exploration. In what is perhaps the most widely read paper on the subject, Rolston (1986) lays out an account of intrinsic value applicable to a wide array of phenomena, terrestrial as well as extraterrestrial, and provides a number of precocious recommendations for the preservation of this value. One of the central tenets of his position is that nature is "projective." By this he means that an unguided tendency of natural processes is the creation of various "projects," which "often culminate in natural kinds, products with wholeness—stars, comets, planets, moons, rocks, mountains, crystals, canyons, seas" (155). These "products with wholeness" are known as *formed integrities*, and are loci of intrinsic value, according to Rolston, because they are "remarkable, memorable . . . so far as they are products of natural formative processes" (156). In general, he claims, "[t]here is value wherever there is positive creativity" (156).

What follows, he insists, is a general duty to preserve formed integrities as well as the processes which give rise to them. To this end he provides six rough guidelines for constraining human activities in space:

6. We should respect "*any natural place spontaneously worthy of a proper name*" (172; emphasis in the original).
7. We should respect "exotic extremes" in nature (173).

---

12. I have in earlier work addressed the views of Sparrow and Lee (Schwartz 2013b), as well as Reiman (Schwartz 2016a). It is perhaps worth stating that I am somewhat less critical of these views than I was in the past, if only because I have come to espouse a more naturalistic attitude concerning the nature of evidence I see as required for justifying, e.g., claims about intrinsic value.

8. We should respect places of historical value (174).
9. We should respect places of "*active and potential creativity*" (176; emphasis in the original).
10. We should respect places of aesthetic value (177).
11. We should respect places of transformative value (177).

Rather than providing a point-by-point exposition and critical discussion of each recommendation, instead I would like to shine a light on those aspects of Rolston's own discussion which I take to be particularly insightful, regardless of our tolerance for the intrinsic value of abiotic entities. In support of (7), he argues that if we value diversity in terrestrial environments, we similarly ought to value diversity in extraterrestrial environments. Moreover, the possibility that certain space environments may be rare or unique should not be ignored, because, he insists, rarity provides a *prima facie* reason for preservation (174). If we accept that we are sometimes justified in placing a higher value on an entity because it is rare or unique (as we often appear to accept), then we should have little difficulty accepting this same practice in application to rare or unique space environments. Rolston also seems aware of the importance of a point raised in Chapter 3 that scientific anomalies, including rare or unusual locations, are ones that can teach us "something about the *nature of things*" (174).

Regarding (8), Rolston encourages us to "preserve those places that have been more eventful than others" (175). Because history "is nowhere even-textured and homogeneous" it follows that some places may possess greater historical value than others (175). He offers locations in which liquid water exists or has existed as being of particular importance, which is consonant with astrobiology's great interest in these locations. However, we should be careful about the way in which we take a place's "eventfulness" to contribute to its historical value. One could easily come to think, especially in light of recommendation (9), that eventfulness requires the presence of *active* or *ongoing* natural processes, and that more or less static environments, e.g., places in which little or no hydrological, geological, or other activity takes place, are less important in this regard. This should be resisted because, in many ways important to planetary science, the *least* active environments are the most fruitful for unraveling one eminently interesting story, *viz.*, the formation and early evolution of the Solar System. As discussed in the previous chapter, the Moon is of particular importance to planetary science precisely because of its lack of endemic geological processes, which makes it an invaluable repository of information about the early evolution of the inner Solar System.

Lastly, with respect to (11), Rolston underlines the potential for the experience of the space environment to alter our perspectives. He draws here on the concept of *transformative value*, which is elaborated more fully in the work of Bryan Norton (1987). An entity has transformative value when experiencing it has the capacity to change our values in ways that we regard as progressive. Consider a person who has never of their own volition read a novel, and who does not consider themself to be living a life that is in any way defective on account of this aspect of their reading habits. Imagine that this person makes a new friend who is an avid reader of science fiction novels. Wanting to know what all of the fuss is about, one day they decide to borrow one of their friend's novels, *Natural History*, by Justina Robson. (Luckily their friend has great taste and owns many novels by British hard science fiction authors!) In the process of reading through the novel they become fascinated with the world-building and visions of the future. After quickly finishing the novel, they ask their friend if they can borrow another (this time an Alastair Reynolds novel, *Revenger*), and before long they become a habitual reader. One day they consider their life prior to their exposure to British hard science fiction, lamenting how parochial and shallow they were compared to the person they have become. That is, in looking back, they appreciate the ways in which their reading experiences have, in their estimation, *improved* their values and their character. In this way science fiction has *transformed* their values, and thus has *transformative value*.

Rolston, in turn, expects that experiences of the space environment will provide analogous opportunities for the transformation of our values:

> We can reduce human provinciality with the diverse provinces of solar-planetary nature . . . . These will prove *radical* places to understand, not merely in the anthropic sense that our *roots* lie there, but in the nonanthropic sense that they *uproot* us from home and force us to grow by assimilating the giddy depths and breadth of being. (1986, 178)

We should perhaps be skeptical about the prospects for experiencing the space environment to "reduce human provinciality." Such an effect is often identified as a benefit of human space exploration (at least for humans who have ventured into space),[13] but such claims have never been tested in a controlled fashion. For instance, it has not been ruled out that those claiming to see the Earth as a singular place, humanity as a singular collection of persons, etc., upon returning from spaceflight, were disposed to think in such ways *before* their respective journeys into space.[14] Still, even if this particular example misfires, that does not mean we are wrong to suspect that the

---

13. See White (2014).

14. *Cf.* Taylor Dark's worry that "[c]laims that space travel will broaden human sensibilities, thus reducing nationalist and chauvinist impulses, are about as convincing as the belief, once widely held, that the view from high-flying airplanes would alter consciousness in a more cosmopolitan direction" (2007, 570). See also Bjørnvig (2013).

experience of the space environment will alter our beliefs and values in important ways. I have noted the tendency for individuals to find value in the things that they study, spend time with, and so on. The relevant lesson here is not that we can identify at the present moment precisely how our values will change, but rather that we have every expectation that they will change, and for the better, if we strive to protect the space environment in ways that allow for humans to experience and come to appreciate it.

One of the most important points Rolston makes is that, given our relative lack of knowledge of and experience with the space environment, we are not especially well positioned to determine, in much detail, what environments, locations, and entities are worth protecting in space:

> Speculating over what places, planets, moons should be designated as nature preserves would be more foolish than for Columbus to have worried over what areas of the New World should be set aside as national parks and wildernesses. All the same, in retrospect, our forefathers would have left us a better New World had they been concerned sooner about preserving what they found there. (Rolston 1986, 171)

This point is put in a way that is perhaps stronger than can be sustained, because our current understanding of the Solar System is much greater than Columbus's understanding of what we today call the North American continent. Indeed, our current understanding of the Solar System is much greater than was Columbus's understanding of North America even at the time of his death. Still, just because we are in an epistemically superior situation when compared to Columbus does not mean we are in an epistemically adequate situation. On a charitable reading, then, Rolston's point is that at present it would be absurd to claim, of any celestial body other than Earth, that we know everything necessary for the construction of informed preservation policies. Best, then, to be more rather than less cautious when it comes to protecting space environments, lest in our ignorance we destroy something of great value.

It is here that Rolston makes most clear that we should not be preoccupied solely with the search for extraterrestrial life. His worry is that too much of an emphasis on life risks blinding us to the value of abiotic entities:

> Banish soon and forever the bias that only habitable places are good ones . . . and all uninhabitable places empty wastes, piles of dull stones, dreary, desolate swirls of gases. To ask what these worlds are *good for* prevents asking whether these worlds are *good* in deeper senses. The class of habitable places is only a subset of the class of valuable places. To fail as functional for Earth-based life is not to fail on form, beauty, spectacular eventfulness. (171)

As a prophylactic against such a bias against non-life, he encourages us to adopt an attitude according to which non-life-harboring worlds "are not places that failed" because:

> [n]ature never fails. Nature only succeeds more or less with its projective integrity . . . . We ought not condemn Mars because it failed to be Earth, although we may value it less than Earth . . . .
>
> Learning to appreciate these alien places for what they are in themselves, not deprecating them for what they failed to be, will provide an ultimate test in nature appreciation. (172)

This is good advice whether or not we agree with Rolston's underlying views on intrinsic value. By approaching our description of space environments in a positive fashion (i.e., with a conscious attempt to articulate what is there in an environment, as opposed to what is not there), we will be more likely to identify places of value, sites of scientific interest, etc., and come to feel compelled to protect these locations, than if we wallowed in disappointment at the absence of life. For instance, describing asteroids as cold dead rocks wafting through space makes them seem hardly worthy of protection, preservation, or study. But describing them as mineralogically heterogeneous, multibillion-year-old clues to the formation of the Solar System suddenly makes them appear more interesting.

Of all individuals this point was not lost on Carl Sagan, who, while lamenting "the sense of disillusionment and disappointment of poets who would prefer to have the moon be unstructured, a kind of Rorschach test in the sky," remarked further that

> if we look closer, we can find poetry there. The magnificent wasteland, as Buzz Aldrin called it, is in fact a record of how worlds are formed. We see the birthing of worlds in the desolation of the lunar landscape, and it applies to every world in the solar system. (As quoted in National Geographic Society 1994, 17)

The central takeaway from Rolston, then, is that we should not proceed in discussion about the scope of and rationale for planetary protection under the presumption that we already know what is interesting, valuable, or worth protecting in the space environment. As I have argued elsewhere (Schwartz 2016a), what we discover and what we learn through the exploration of space will affect what we find interesting and what we value. Similarly, what we find interesting and what we value will affect what we decide is worth exploring and studying. We should for these reasons adopt an attitude that is more rather than less tolerant about sources of value. And, it is worth repeating, we need not agree with Rolston that the reason for doing this is because of a duty to respect loci of intrinsic value (formed integrities) wherever they appear. After all, other values are at risk, including values associated with the use of space for the advancement of scientific understanding. Importantly, this understanding should not be restricted to the potential insights of astrobiology but should include the perspectives of space science, broadly speaking. Only thus can we properly engage in the sort of positive description and appreciation of the space environment for what it is in itself. While we could get carried away and become oversensitized to the space environment, for the foreseeable future the opposite problem is much more likely to occur—what we need for the moment is a growing rather than attenuating appreciation of the space environment.

Milligan

Some might balk at Rolston's position, if for no other reason than they find his relatively permissive, non-biological account of intrinsic value to be too great of a metaphysical commitment. I hope that does not diminish the significance of the insights he provides, but in any case, it is worth exploring an alternative approach, *viz.*, Tony Milligan's character-oriented perspective on duties to (or involving) the space environment. For Milligan, what is important is that we are a part of humanity, i.e., that we belong to human communities and human cultures. It follows from these various community and cultural memberships that we share certain values and obligations, with these in turn being constitutive of what it is to live a "life like ours." He thinks it is unlikely that our shared values and obligations could ever be reduced to simple maxims (e.g., Kant's categorical imperative) but that instead,

> there is a wide range of ethically-sensitive concepts which play a role in our best descriptions, often without drawing attention to themselves: betrayal, fairness, cruelty, humility, courage and humanity itself. The list may not be a short one. (Milligan 2015a, 47)

However, the precise set of values and obligations we share with others is contingent and in many ways depends on the wider environmental conditions in which we live our lives. Thus, if we someday begin to spend most of our lives in heretofore novel circumstances, for instance, in space settlements, we should not be surprised to see our values and obligations adapt to better fit the maintenance of human societies in these new settings. Although some values and obligations are likely to remain (a prohibition on murder, for example), nevertheless:

> [t]here may always be room for another concept or a modified concept, one which helps us to understand what it is to be human, to be bound to other humans and

to other creatures, or which helps us to understand what it is to be connected in deep ways with some place or other: a village, a country or home planet. (48)

The idea that it is an important human value that we connect deeply with *places* is one that has implications for planetary protection, and Milligan is skeptical that protection for the sake of science will "protect all that needs to be protected" (2015b, 129).

What then is involved in connecting deeply with a place, whether it is one in which we reside, merely visit, or observe from afar? For Milligan this comes to a recognition and appreciation of the *integrity* of that place. His conception of integrity is not dissimilar from Rolston's idea of formed integrity, though Milligan does not claim to be engaging in any kind of moral metaphysics. Instead, he intends to capture a sort of manner of *speaking* or *thinking* about places (including extraterrestrial places) that validates our tendency to value places "in ways that go beyond any belief that we or others can gain something useful from them" (2018, 3). In this respect he notes that "at least sometimes we care about places because they are the places that they are, because they are unique and enhance the diversity of the world, because of their special history, and for similarly non-instrumental reasons" (8).

Milligan's notion of integrity, then, is a sort of catch-all or shorthand for the breadth of non-instrumental reasons we might value a place. Such language is not empty, he claims, because "if we say that 'the integrity of x' has been compromised, no-one will be confused about what is meant" (10). More fully:

> A place with integrity will have a distinctive, unique, or near-unique structure or composition. This structure or composition will be present because of a unique history and will contribute to diversity. (10)

Furthermore, places can admit of greater or lesser degrees of integrity. Clear examples of places with greater degrees of integrity are planets and moons, as well as large asteroids and planetoids such as Ceres and Vesta (11). Clear examples of places lacking integrity are "commonplace rocks and the over-whelming majority of objects in the asteroid belt," with marginal cases including "comets and the interstellar medium" (11). Recognizing, valuing, and respecting the integrity of places is compatible with the alteration and use of these places, at least to an extent. Milligan grants that deliberate changes comparable to naturally occurring changes pose little threat to integrity. However, matters are different when considering "cases that involve more extensive alterations or structural damage," including the strip-mining of the Moon for ^{3}He (11). The underlying principle here seems to be that, the greater the threat that an action poses to the integrity of a place, the more difficult it is to justify that action. On this picture, it is entirely possible that a threat of complete destruction of the integrity of a place is justifiable, if that place was of little integrity to begin with, and if that destruction served an important enough purpose. Meanwhile, there may be no human purpose important enough to justify extensively undermining the integrity of a place like the Moon, which possesses a great deal of integrity.

As with Rolston, I think there is merit to Milligan's suggestion that we remain vigilant for non-instrumental reasons for valuing places. I am concerned somewhat about the intentionally nebulous nature of his notion of integrity, however. Consider the case of the Moon. Its integrity seems to be an agglomeration of a variety of its characteristics: its beauty; its surface topography and composition; its deeper structure and geological features; etc. Are we to understand its integrity as the corporate whole of these features? Or would it be best to think of the Moon as having several distinct integrities? Should we speak primarily of the integrity of the Moon *simpliciter*? Or speak instead of the integrity of its beauty; the integrity of its surface; the integrity of its interior; etc.? Already, then, we have a reason to doubt that "no-one will be confused about what is meant" (10) when discussing violations of the Moon's integrity.

While the choice is not an exclusive one, Milligan seems comfortable speaking about the integrity of *components* of the Moon (11). Thus, compromising the integrity of Tycho Crater (Fig. 4.1) would not be tantamount to compromising the integrity of the Moon. Still, even at this level of analysis, there are multiple ways in which we might speak about the integrity of Tycho Crater: its beauty; its uniqueness; its topography; its composition; its history; etc. Compromising the compositional integrity of Tycho Crater might be compatible with the preservation of its beauty, topography, and uniqueness, for instance if a dedicated "lunar landscaper" replaced its surface regolith with phenomenally indistinguishable materials harvested elsewhere from the lunar surface. This raises questions familiar to restoration ecologists—how does the value of a replica compare with the value of the original? Can environmental destruction be justified if it is followed by restoration of the damaged environment? But these are somewhat orthogonal to point I am raising, which is that there may be no such thing as the integrity of Tycho Crater (or any other place) *simpliciter*. Rather, there may only be various *integrities* associated with Tycho Crater. This is not an objection to Milligan so much as making plain that acknowledging the integrity of places will not necessarily simplify the deliberative process. Indeed, it raises the prospect of complicating it significantly, even if the complications are welcome and long overdue.

A second point of concern is with Milligan's prediction that protection for the sake of science is likely to be inadequate, all things considered. Although this almost certainly seems true when our attention is restricted to places of interest in the search for extraterrestrial life, nevertheless this point loses

Figure 4.1 Tycho Crater is a lunar impact crater in the southern lunar highlands, and would seem to qualify as a place with integrity—but in what way?
Credit Line: NASA Goddard/Arizona State University.

much of its force when we expand protection to *all* places of potential scientific interest in space. Science has yet to disclaim interest in any particular celestial body, and thus, plausibly, the entirety of the Solar System could warrant protection in the interest of science. Still, the set of places of integrity and the set of places of scientific interest may not be coextensive, and the former are, in a sense, more permanent. If we have an obligation to protect a place because of its integrity, that obligation remains should its scientific novelty ever wear off. Thus, an ethic like Milligan's may increase in relevance as humans come closer to attempting more destructive uses of space environments. This should not detain us from appreciating that, for the time being, scientific values are especially salient in decisions about planetary protection. Furthermore, science has an important contributory value: The understanding gained through the studious scientific examination of the space environment will increase our awareness of which places possess which kinds of integrities and to which degrees. Milligan's ethic is largely impracticable if it does not reserve an important place for scientific interests.

The Contribution of the Value of Science

Regardless of one's sympathies for claims about the intrinsic value of the space environment, or about our obligations to respect and preserve the integrity of places, we ought to recognize a variety of scientific values when deliberating about planetary protection. As I argued in Chapters 2 and 3, the advancement of scientific understanding is both intrinsically and instrumentally valuable, and we have a *prima facie* duty to realize this value—a duty that in practice is not overridden and so in fact compels action on our part. Because the scientific exploration and examination of the space environment promises extensive contributions to our understanding of the universe, there exists a corresponding obligation to carry out the scientific exploration of the space environment. Since this exploration will be most effective through the examination of *pristine* environments, there exists a corollary obligation to protect these environments to avoid diminishing the effectiveness of our exploratory activities. On this point the great expenses associated with space exploration should not go unappreciated. This and other barriers to planetary exploration make it important to maximize the effectiveness of missions, that is, to maximize the scientific return on the very limited funding granted to space exploration missions. Contamination or disruption of target destinations dramatically increases the potential for ambiguity in the conclusions drawn from the exploration of these locations. A scientific mission sent to an environment that had been previously contaminated or disrupted would be much more difficult to justify than a similar mission sent to a pristine environment, at least if the goal of the mission is to learn about that environment in its natural state. Why spend potentially billions of dollars on a space mission if you cannot rule out that any potentially interesting features or anomalies you discovered were actually the result of prior contamination? That is not to say that there is no reason to be interested in learning about the effects of our actions on the space environment (or about the effects of the space environment on discarded hardware). It is just to point out that one very important kind of goal should not be jeopardized haphazardly or nonchalantly, *viz.*, the study of space environments undisturbed by human spaceflight activities.

Though this is an instrumental argument of sorts—protection of the space environment is warranted not for the sake of the space environment itself but for some further purpose—this further purpose is the promotion of a nontrivial, intrinsic good: the advancement of scientific understanding. Moreover, while good in itself, the advancement of scientific understanding also has a contributory value under views like Rolston's and Milligan's that ask us to appreciate space environments for what they are in themselves, as opposed to denigrating them for what they lack. Such appreciation requires an underlying understanding of what these space environments are like—an understanding that can only be provided through scientific exploration. *Science provides the empirical foundation of any practicable space ethic!* This, of course, raises the question as to whether any sites of interest to science are at genuine risk, aside from those implicated in the search for evidence of extraterrestrial life (which already fall under the scope of existing planetary protection policies).

The potential for the space environment to be damaged or contaminated in ways deleterious to science has been noted by Iván Almár. The trouble, he claims, is that

> in case planetary explorers do not fully address the environmental consequences of their activity and do not protect the pristine surface and subsurface of celestial bodies, all essential *in situ* evidence on the origin and evolution of planets, asteroids and satellites will be denied [to] future generations of astronomers. (2002, 1577)

The appropriate level of concern is partly a function of the level of geological activity in an environment. For instance, on "fast changing surfaces" like the surface of Io, which is exceptionally volcanically active, "there is generally less danger that human interaction would destroy something which otherwise would stay unaltered for billions of years" (Almár 2010, 28). But "very old, non-variable surfaces such as the Moon, Mercury, and several satellites and asteroids, are absolutely vulnerable and their scientific value would be irrevocably damaged if a large-scale intervention occurs" (28). William Kramer highlights the nature of the problem this would pose to future scientific examination:

> Without an environmental assessment and vetting process, it may be increasingly difficult to differentiate human from natural features. Does the presence of a ridge of rocks at the mouth of a valley indicate some ancient moraine or other geologic process, or is it the result of a mineral prospecting mission? Were rocks fractured due to geologic factors, or were they shattered as part of a research project decades earlier? (2014, 218)

Kramer provides a minimal recommendation, which is to develop environmental impact assessments for spaceflight activities in order to provide future explorers with information about the extent to which, e.g., planetary surfaces have been disturbed by human activities. Almár, meanwhile, suggests that more extensive work is required, and that scientific stakeholders should "survey and evaluate all existing planetary environments with regard to their scientific value . . . sensitivity to artificial interference, [and] difficulty or ease of access by planetary missions" in order to identify which environments might welcome human explorers, which environments should be explored exclusively by robots, and which environments should be protected completely (2002, 1579). Of course, given our poverty of understanding when it comes to geological processes on other planets, we might need to do quite a lot of scientific exploration before we can begin to form a coherent and actionable understanding not only of the potential impacts of human activity in the space environment, but also of which aspects of the space environment are scientifically valuable in which ways.

As discussed in the previous chapter, sites of legitimate scientific interest exist at virtually every location in the solar system, from the inner planets and the Main Belt asteroids, to the gas and ice giants and their satellites, and to the outer satellites, Kuiper Belt, and cometary nuclei of the Oort Cloud. Since science has yet to disclaim interest in any particular celestial body, a precautionary principle is justified, in which *a space environment should be presumed to be of scientific interest until proven otherwise*.[15] Nevertheless, in practice it will be difficult to maintain that the entire Solar System should be protected as a scientific preserve, and so it would be helpful to see Almár's entreaty given more consideration. Some small steps have been made, including Cockell and Gerda Horneck's idea of creating "planetary parks" (Cockell and Horneck 2004; 2006). More recently, Jack Matthews and Sean McMahon have adumbrated tentative criteria for identifying sites of scientific value, or what they call "exogeosites." These are places that may

- serve as a representative reference "type" exemplifying a class, or be among very few known specimens of that class on a particular celestial body;
- be a particularly well studied feature for which key data are uniquely well constrained, and thus an important standard reference point even if not remarkable for any other reason;
- be an excellent resource for scientific education and training, perhaps by virtue of being well understood by experts but challenging to interpret for novices in the field;
- be the subject of ongoing research of particular importance or significant controversy;
- possess clear potential to drive future research once present technological limitations have been overcome, e.g., in the miniaturization of isotope mass spectrometers for Mars missions. (2018, 56)

15. *Cf.* Stoner (2017) and Gottlieb (2019).

In advance of extensive in situ exploration, it may be difficult to determine which locations serve as representatives of "types" of space objects, or which locations are home to rare or unusual objects or phenomena. This again recommends a cautionary approach and suggests that activities such as asteroid mining should not be undertaken until we know enough about the nature, composition, and distribution of asteroids to guarantee that such exploitative acts would not destroy an object of unique scientific significance. Of this, more in Chapter 5.

There are, then, legitimate scientific reasons motivating planetary protection, broadly speaking. It is also worth reemphasizing the contributory value of scientific study to considerations like those raised by Rolston and Milligan, especially when it comes to the aesthetic appreciation of the space environment. As McMahon laments, our sense of what is or might be beautiful about Mars is not always mediated by an awareness of what Mars really is like, but often through scientifically uninformed depictions, for instance in novels, film, and television. As Rolston and Milligan would be quick to point out, an appreciation of Mars's beauty based on inaccurate or embellished depictions could result in our becoming disenchanted with Mars as it really is. But if McMahon's own experience as a planetary scientist is any guide, a studious examination of Mars as it actually is (for instance, in Fig. 4.2) can provide a window into a genuine aesthetic appreciation of Mars:

> To the extent that a serious, true-to-nature aesthetic engagement with Mars is possible today, I think it must involve, firstly, close and open-minded attention to the images of Martian landscapes currently available for study and, secondly, the imaginative contemplation of these images in light of the known facts about Mars. The character and quality of aesthetic experience, especially of natural scenery, is often strongly determined by contextual knowledge. (2016, 216)

By "contextual knowledge" he means that aesthetic appreciation is not innate but instead is reflective of the knowledge one has about a given object. Those with more knowledge about an artwork, its history, its creator, etc., are often capable of higher levels of aesthetic appreciation than those with less knowledge of these topics. Similarly, those with greater knowledge and understanding of Mars should be capable of more discerning taste when it comes to appreciating distinctively Martian forms of beauty. In his own case, McMahon remarks that "by studying Mars I have come to recognize beauty of an unearthly kind in its pale colours and pure, ancient landforms" and that "the destruction of this kind of beauty would seem to be a terrible loss" (217). In addition to planetary scientists, we might also solicit the perspectives of landscape artists, novelists, poets, musicians, and other creative agents who might impart to us novel and enlightening ways of appreciating places like Mars for

Figure 4.2 This image of Bunge Crater Dunes on Mars could easily be confused for abstract art.
Credit Line: NASA/JPL-Caltech/Arizona State University.

what they really are. Crucially, such opportunities for creative expression would be lost forever if we deprived ourselves of opportunities to generate an underlying understanding of Mars as it really is.

INTERSTELLAR IMPLICATIONS

We may become interstellar explorers sooner than many realize. Although *Voyager 1* has technically entered interstellar space, and *Voyager 2* is expected to do so in 2020 (with *Pioneer 10* and *11* to follow, along with *New Horizons*), nevertheless neither probe is expected to approach another star system for tens of thousands of years. Meanwhile, the relatively rapid progress of the Breakthrough Starshot initiative, which hopes to send gram-scale probes to the Centauri system using laser propulsion, might be realized on the order of decades. Younger individuals alive today could witness the first images of an

The Value of Science in Space Exploration

extrasolar planet taken from *within* its home system. Now, therefore, would be an opportune time to initiate serious discussion about planetary protection for interstellar missions. As things stand currently, COSPAR's policies only place requirements on missions to destinations within the Solar System, and so any deliberate attempt to send a probe to another star system would not enjoin any protection measures. And while a gram-scale probe, thoroughly sterilized by the interstellar radiation environment (Cockell 2008a), would pose scant chance of biologically contaminating its target system, there are more ambitious (though much more tentative) proposals afoot.

Claudius Gros's *Genesis Project* (2016; 2019) studied the feasibility of using a Starshot-style mission to deliberately send life to another star system. On such a "life-seeding" mission the goal might be either to spread life to elsewhere in the universe or to prepare an exobiosphere for later human settlement. Another mission concept with a similar set of goals would be to send self-replicating "von Neumann"-style machines to another star system (Freitas 1980). Upon arrival, these machines would manufacture copies of themselves using in situ resources, possibly sending these copies to further star systems, or directing them to prepare the system for future human settlement. Proposals such as these are incredibly improvident from the point of view of anyone concerned with protecting opportunities for scientific study. To borrow an expression of Sagan's, an entirely unexplored star system would be a treasure beyond assessing, a treasure that would be lost entirely if a life-seeding or self-replicating probe mission led to the destruction within its target system of huge tracts of in situ evidence of the history and nature of the objects it encountered. Given how much might be lost, it is imperative to restrict early exploration of other star systems to purely observational missions, i.e., missions which can be implemented successfully without any in situ resource consumption (other than, e.g., the collection of solar energy).[16]

Even though probes may be sent out within decades, in the absence of any game-changing technological breakthroughs, human interstellar exploration remains centuries, if not millennia, beyond our reach. There is no good reason for us today or in the near future to send out interstellar settlement precursor missions. Doing this is incredibly unlikely to improve the chances of human survival, even over the incredibly long term. This means that we have centuries (at the very least) during which we can reserve other star systems for scientific study. It strains credulity to maintain that our understanding of our own Solar System has progressed to the point that we now know everything worth knowing about what is worth protecting in it. It would be absolute nonsense to claim that we now know enough about any other star system to say what is and what is not worth protecting within it. The protection of interstellar

16. See Schwartz (2019b) for further discussion.

science, then, provides sufficient motivation for the construction of protection policies for interstellar missions. Though some familiar protocols may be appropriate (e.g., we may require of Starshot-style missions that they implement trajectory biasing to avoid impacting objects within target systems), this topic should be approached as planetary protection is currently, i.e., with protocols updated regularly and which reflect advances in technology as well as in our understanding of target environments and of our potential for disrupting those environments.

CHAPTER 5

The Need for Forbearance in Space Resource Exploitation

Spaceflight activities are not limited to science missions. There are many commercial and national security services provided from space, though for the moment these activities are confined to various Earth orbits. However, there is growing commercial interest in moving beyond Earth orbit to the Moon and near-Earth objects (NEOs) where many potentially valuable resources await, such as: platinum, nickel, titanium, iron, aluminum, and water. Shackleton Energy Company, for instance, hopes to exploit the water ice deposits in lunar polar craters. This water could be processed into liquid hydrogen and oxygen, which together are a useful rocket propellant. Moon Express intends to use robots to prospect the Moon in preparation for the future exploitation of its resources. Meanwhile, Planetary Resources, Inc. hopes to profit from mining near-Earth asteroids.

What explains the commercial interest in space resource exploitation? After all, the resources just mentioned can be acquired much more easily and with considerably less expense from terrestrial sources. At the same time, spaceflight is incredibly expensive. The cost of reaching low-Earth orbit (LEO) remains in the range of 2,000 to 10,000+ USD/kg, with greater expenses associated with higher orbits and further destinations. The biggest factors contributing to these high costs are the demands placed on the design, construction, and testing of launch vehicles and payloads. With regard to the former, one of the requirements is that the vehicle be capable of accelerating its payload to orbital velocities (around 7.8 km/s). Launching from sea level, and due to atmospheric drag and gravity losses, this requires a minimum Δv (change in velocity or "Delta-V") cost (also known as a Δv budget) of 9.4 km/s (over 21,000 mph). Reaching a higher orbit, or traveling to other destinations, such as the Moon or Mars, requires more energy yet. Precisely how much propellant is required depends on the design of the launch vehicle, the efficiency of its engines, the energy potential (specific impulse) of its propellants, and the mass of the payload as well as of the launch vehicle itself (though the cost of propellant tends to be a minor factor when compared to launch vehicle costs, payload costs, and insurance). Rather than launching massive payloads, if one could instead construct and fuel spacecraft from resources already in space, then one could dramatically reduce the cost of spaceflight activities. It is hoped that mining the Moon and NEOs can provide the spaceflight industry with resources in situ and at a considerable cost savings compared to the value-added cost of resources launched from Earth.[1]

Unfortunately for space mining enthusiasts, the legal situation surrounding space resource exploitation is uncertain. A key issue is the Outer Space Treaty, ratified by all spacefaring nations, which prohibits national appropriation of the space environment:

> Outer space, including the moon and other celestial bodies, is not subject to national appropriation by claim of sovereignty, by means of use or occupation, or by any other means. (Article II)

In conjunction with Article VI, which identifies states as legally responsible for any actions taken during a mission launched from their soil, it is generally understood that Article II's prohibition on national appropriation precludes any claim to private ownership of the Moon, asteroids, or other celestial bodies. Though recent legislation from the United States and Luxembourg attempts to circumvent this (and is discussed briefly below), the legality of commercial exploitation of space resources remains unclear. The nascent space mining industry is understandably interested in pushing for changes in the legal environment in order to attain the certainty it perceives as a precondition for taking the immense monetary risks associated with lunar and asteroid exploitation. After all, no business could be expected to spend hundreds of millions of dollars or more extracting resources that they would not have the legal right to sell on the global market.

The question I would like to address in this chapter is whether we ought to cede to the wishes of the space mining industry by implementing a legal regime that allows for, or perhaps even encourages, the commercial mining of the Moon

---

1. For instance, the components of one common fuel mixture, RP-1 (a refined form of kerosene) and LOX (liquid oxygen) cost about 30.60 USD/kg and 2.40 USD/kg, respectively (Defense Logistics Agency, Standard Prices, 2017 Standard Prices for Aerospace Products, available at: http://www.dla.mil/Energy/Business/StandardPrices.aspx (accessed 28 July 2018)). At 5,000 USD/kg to LEO, these propellants are worth around 150,000 USD/kg and 12,000 USD/kg! Thus it might become profitable to harvest lunar and NEO water for LOX production, if this can be done for less than 12,000 USD/kg.

and asteroids. While this question concerns *whether* to adopt laws and regulatory frameworks of various kinds, I shall in this chapter view legal questions as secondary to ethical questions. My concern, then, is whether commercial space mining would be ethically praiseworthy, neutral, or blameworthy. And while I grant that there is a *prima facie* case supporting an obligation to exploit space resources for commercial gain, it is one that I shall argue is overridden by other, conflicting and more pressing obligations. The position I shall defend here is one according to which space mining should not be prioritized, and thus, should not be encouraged through the establishment of mining-friendly legal instruments.

In The Case for Space Resource Exploitation I lay out the standard case for the exploitation of space resources and for implementing an exploitation-friendly legal regime. There are two basic motivating considerations: First, regime change is needed because it would enable the exploitation of resources that could be used to mitigate terrestrial resource depletion (and its various human and environmental costs). Second, regime change is needed because it would enable the exploitation of resources that could be used to improve overall human well-being. Those promulgating such arguments tend to view space as containing abundant, limitless resources of all kinds, obviating the need for any limits to human consumption. But as I argue in The Scarcity of Space Resources, this is not entirely true. When our attention is restricted to nearby, *easily accessible* destinations in space, resources like water and platinum-group metals are relatively scarce. This makes it unlikely that space mining would help to stave off terrestrial resource depletion in any significant way over the near future. It is equally doubtful that the exploitation of space resources would benefit all of humanity. As I argue in Space Resources: Benefits for Humanity?, the liberal, laissez faire approach to regulation favored by space mining advocates would only serve to exacerbate existing inequalities, both at the domestic and international levels.

The upshot is that the ethical case presented by space mining advocates is not strong. As I explore in Implications for Future Policy, this affects our approach to the construction of ethically sound space exploitation policies. An important point of this discussion is that most existing regulatory frameworks and proposals aim to legislate *access to* space and its resources, where the goal is to ensure some kind of equal or equitable *opportunity* to exploit the resources of space. However, certain of these resources are so practically scarce as to make it impossible for every interested party to acquire a useful share. For this reason I argue that future regulation must legislate the *use* of space resources. If there is not enough for everyone to do what they want with, e.g., lunar polar water, then we must adopt a system in which lunar polar water is not wasted on trivial fancies (e.g., swimming pools for opulent lunar hotels) but is instead reserved for activities that realize greater ethical goods (e.g., fuel for science missions). The clear moral justification in favor of space science marks it as the most ethically salient stakeholder when it comes to the use of the space environment, including the use of any space environments of interest to the space mining industry. Given the likelihood of conflicts between scientific investigation and exploitation and between consumption of resources for scientific purposes and consumption of resources for other purposes, we ought only to pursue regulatory mechanisms that prioritize the former of each pair over the latter.

THE CASE FOR SPACE RESOURCE EXPLOITATION

Those advocating for space resource exploitation often take themselves to be advancing moral arguments that establish an obligation to exploit space resources. One claim is that space resource exploitation is needed to mitigate the worst that terrestrial resource depletion might bring our way. As Lawrence Cooper has argued, any duty on our part "to protect the environment can be interpreted as an imperative to colonize space and exploit its resources" because given the now seven billion plus (and growing) human population, "space beckons with the promise of unlimited resources to augment the ever-shrinking terrestrial resources and relieve the stress humans put on the environment" (2003, 115). Thus, if we make the initial down-payments associated with developing a space mining infrastructure, we may never again have to worry about running out of most, if not all mineral resources. The exploitation of space resources even promises to solve our energy needs. The lunar regolith contains numerous solar implanted volatiles, including trace quantities of the helium isotope ^{3}He, which would be an especially valuable fuel for fusion power. It has been suggested that the quantity of ^{3}He available on the Moon is enough "to meet the world's energy demands for thousands of years, even if we don't develop additional alternative clean energy sources" (Matloff, Bangs, and Johnson 2014, 115). Meanwhile, the construction of many thousands of photovoltaic satellites out of lunar or asteroid material could enable the development of a beamed solar power grid similarly capable of meeting human energy needs for centuries to come (Kennedy, Roy, and Fields 2013).

Beyond freeing us from the worst that terrestrial resource depletion might bring, the exploitation of space resources would have beneficial knock-on effects. If, for instance, the majority of industrial activity is moved to space, rather than left on the surface of the Earth, we will significantly reduce terrestrial pollution and all of its deleterious effects on terrestrial ecosystems (Matloff, Bangs, and Johnson 2014, 12). Moreover, the opening up of new markets and vast stores of resources promises new wealth, with all of the opportunities such wealth brings. According to John Hickman, accessing the resources of space will improve international relations:

> [L]iving solely on the Earth has not made our species more responsible. Indeed, it might produce more irresponsibility because problems of global governance

become zero-sum struggles. The more politically-acceptable answer is that additional power resources available to spacefaring states, because of their annexations of extraterrestrial territories, would give them the means to better support global governance. (2010, 69)

Hickman's prediction is that if terrestrial nations were no longer forced to battle over the comparably limited, finite resources of Earth, then we would see an easing of the tensions between nations regarding energy sources, for instance. (Of course, if a smaller nation was currently profiting from selling rare resources to wealthier states, what happens when this nation is forced to compete with space-based producers?) Potential benefits are not limited just to states, but are also available to individuals. As Daniel Pilchman argues, exploiting space resources could improve the lives of individuals:

> [O]ne might think that asteroid mining actually represents one of the greatest opportunities for human flourishing in history. Not only does asteroid mining promise wealth, which is essential for the instantiation of virtues like generosity and magnificence, it is a chance for humans to be creative, inventive and explorative. The inherent risks occasion bravery. The ambitious scale of the project makes it an opportunity to achieve glory. These are all human virtues, arguably constitutive of the best human life. Not only is asteroid mining compatible with flourishing, it creates a new way for humans to achieve it. (2015, 144)

For some, the mining of asteroids and the Moon might provide an opportunity for achieving eudaimonia.[2] And if those who benefit directly from this mining are generous enough, either of their own volition or through their contributions to tax revenues, their wealth might contribute to the well-being of the rest of humanity.

These and other rationales for space resource exploitation provide a moral basis for seeking regulatory change. As I have mentioned already, the extant legal environment, via the Outer Space Treaty, is not perceived as amenable to commercial exploitation of space resources, largely because of Article II's prohibition on national (and thus, private) ownership claims of territory in space. Interestingly, the 1979 U.N. Agreement Governing the Activities of States on the Moon and Other Celestial Bodies, or Moon Agreement, did provide a legal basis for exploitation of lunar and other space resources. However, this agreement also stated that "[t]he moon and its natural resources are the common heritage of mankind" (Article XI, Par. 1) and required that the exploitation of lunar resources be conducted under the auspices of an international

---

2. The concept of eudaimonia is historically associated with Aristotle's virtue ethics, and refers to the possession of character traits which enable one to lead a life of virtue or excellence; see Kraut (2018) for elaboration and discussion.

regime (Article XI, Par. 5) responsible for the "rational management" of those resources (Article XI, Par. 11) as well as for providing an "equitable sharing by all States parties"

> in the benefits derived from those resources, whereby the interests and needs of the developing countries, as well as the efforts of those countries which have contributed either directly or indirectly to the exploration of the moon, shall be given special consideration. (Article XI, Par. 11)

Finding these provisions unduly restrictive, few states signed the treaty, and even fewer ratified it, with no major space powers among either group.[3] And, despite attempts to argue that private exploitation of lunar resources is permissible under the Moon Agreement prior to the establishment of the international regime (Brearly 2006), legal scholars generally agree that the Moon Agreement, like the Outer Space Treaty, is similarly prohibitive of national and private appropriation of the Moon.

The Moon Agreement, if widely signed and ratified, and if adequate support were given to the development of the international regime described in Article XI, would appear to provide humanity with a path for realizing the kinds of benefits adumbrated above. Why, then, did it not attract more signatories? As many commentators and legal scholars have noted, the sharing of benefits under the Moon Agreement was not well received by the major space powers, and nor has it been well received by representatives of the nascent space mining industry. Why invest considerable capital in space mining if, instead of being able to keep most or all of the profits, one is forced to share the fruits of one's labor with everyone else? The adoption of the Moon Agreement would, most space mining advocates agree, result in the continued stagnation of space exploitation efforts. Thus, Ram Jakhu and Maria Buzdugan argue that the Moon Agreement's "communitarian property system will inhibit economic development and leave exploration and settlement in the realm of governments, who cannot always afford to undertake such activities," and therefore we need "[a] strong private property regime" which could "encourage commercialization and settlement of outer space" (2008, 210).[4] This point is put even more starkly (if not also caricatured) by Virgiliu Pop, who claims that an "egalitarian regime" like the Moon Agreement,

---

3. As of 2018, states signing and ratifying the treaty include Armenia, Australia, Austria, Belgium, Chile, Kazakhstan, Kuwait, Lebanon, Mexico, Morocco, Netherlands, Pakistan, Peru, Philippines, Saudi Arabia, Turkey, Uruguay, and Venezuela. States signing but not ratifying include France, Guatemala, India, and Romania.

4. See also (Solomon 2008), who advocates for abrogating the Moon Agreement because it is "unacceptable to space-faring nations in light of the risks involved in getting to the Moon and extracting its resources" (111).

which not only prohibits private landed property but also calls for an 'equi-table' *jus fruendi* offers no incentive to exploit the extraterrestrial ore when the investors have to share the benefits with free riders who believe in a culture of entitlement. Together with Locke, we believe that law ought to provide for the "industrious and rational" and not for the "quarrelsome and contentious." (Pop 2012, 560)

In place of the Moon Agreement, many representing private sector interests would prefer a "well-defined and minimally regulated legislative regime" so as to encourage and protect the nascent space mining industry (Solomon 2008, 2).[5] This generally is taken to require a clarification, revision, or abrogation of the Outer Space Treaty.

Both the United States and Luxembourg are currently testing the wa-ters with national legislation designed to encourage space mining. In 2015 the United States passed the Commercial Space Launch Competitiveness Act (CSLCA). According to Title IV of this act, the U.S. government entitles U.S. citizens "engaged in commercial recovery of an asteroid resource or a space resource"

to any asteroid resource or space resource obtained, including to possess, own, transport, use, and sell the asteroid resource or space resource obtained in ac-cordance with applicable law, including the international obligations of the United States. (Sec. 51303)

The CSLCA defines "asteroid resource" as "a space resource found on or within a single asteroid" and "space resource" as "an abiotic resource in situ in outer space" which includes water and minerals (Sec. 51301). In this way the CSLCA hopes to circumvent Article II of the Outer Space Treaty as it does not en-title U.S. citizens to claim ownership of any celestial *body*, but instead of the *resources found on or within* celestial bodies. Whether this is genuinely per-missible under the Outer Space Treaty is a matter of dispute,[6] and it is also questionable whether any resources acquired under the auspices of the CSLCA would be transferable on the global market (von der Dunk 2018). Thus the CSLCA, along with the similar Luxembourg Law on Use of Resources in Space, will provide interesting tests of the robustness of the Outer Space Treaty, should any U.S. or Luxembourg firm actually engage in space mining.

Details of extant national legislation aside, the question remains as to whether such legislation is ethically sound and whether something resembling

5. See also (Lee 2012) and (White 2002).
6. For contrasting analyses see Basu and Kurlekar (2016), De Man (2019), and Perry (2017).

it should be implemented at the international level. The core argument for doing this can be summarized as follows:

1. We have obligations to, e.g., mitigate the risks associated with terrestrial resource depletion, and improve the material well-being of humanity.
2. Certain forms of space resource exploitation will help us to satisfy these obligations, and may even be necessary for doing so.
3. Thus, *ceteris paribus*, we have an obligation to exploit those space resources which will help satisfy the obligations mentioned in (1).
4. The extant regulatory environment is prohibitive of exploitation of space resources.

———

5. Therefore, we have an obligation to implement regulations more supportive of space resource exploitation.

In what follows, I hope to undermine this core argument by casting doubt on premise (2) and by placing stress on the *ceteris paribus* clause of (3). I raise doubts about (2) in The Scarcity of Space Resources by arguing that space resources are scarce in critical ways and thus incapable of providing humanity with the kind of influx of new resources that would help mitigate resource depletion on a significant scale. I raise further doubts about (2) in Space Resources: Benefits for Humanity?, where I argue that the exploitation of space resources is likely to benefit only the already wealthy, and thus, will not be of benefit to the majority of humanity. And with respect to (3), I argue in Implications for Future Policy that any change in the legal environment should be towards a system that is more rather than less capable of promoting and protecting *scientific* interests in space, since scientific exploration is subject to a comparatively stronger ethical justification, all things considered.

THE SCARCITY OF SPACE RESOURCES

There are staggeringly vast amounts of resources available in the Solar System. John Lewis has estimated that through the use of asteroid material alone the Solar System could sustain on the order of *quadrillions* of humans (2015, 103). However, not all destinations in the Solar System are equally attractive. In so far as one's goal is to provide humans on or near to Earth with resources at the lowest possible cost, then one will quickly encounter many practical limitations on the quantity of resources available. The basic problem is that most space resources are so energetically distant that we would be better served by seeking terrestrial sources (or paying for terrestrial resources to be launched into space). This means that for space resource exploitation to be economically viable, we must undertake exploitation only at destinations with minimal Δv

budgets. A reasonable baseline to work with here is lunar escape velocity, i.e., the Δv budget associated with launching material from the lunar surface to beyond the Moon's sphere of influence. This value is a Δv of 2.37 km/s, and provides for a useful comparison of available resources from the two "nearby" resources: the Moon and NEOs.[7] As discussed below, several vital resources are available at this budget, including water and platinum-group metals, although in very limited quantities. This vitiates perspectives according to which space resources are viewed as limitless.

Lunar Resources

Ian Crawford (2015) provides an overview of what is known about the availability of lunar resources—based on studies of: orbital surveys, samples returned from the Moon, as well as lunar meteorites. Though poor in organics, the Moon nevertheless contains significant quantities of solar implanted volatiles, metals, silicates, and water. The Moon also has areas in its north and south poles with high illumination conditions, the so-called "peaks of eternal light," which would be attractive locations for collecting solar energy.

Solar Implanted Volatiles

The lunar surface is comprised mostly of regolith, a fine dust that is the result of billions of years of impact events. This material is rather good at collecting various elements carried along in the solar wind, and which can be extracted by degassing the regolith. Hydrogen is present in the regolith at a concentration of 46 ± 16 parts per million (or 76 g/m^3) (142). Nitrogen is present at a concentration of 81 ± 37 parts per million (or 135 g/m^3) (143). But far and away the most valuable commodity here would be the helium isotope ^{3}He, which is present at a concentration of 0.0042 ± 0.0034 parts per million (or 0.007 g/m^3), though it may be present in higher concentrations (between 10 and 20 parts per billion) in certain lowland regions, such as Oceanus Procellarum and Mare Tranquillitatis (143–145). Given the incredibly low concentrations associated with ^{3}He, exceptionally large tracts of regolith would have to be processed in order to produce even very small quantities of the helium

---

7. Note that the Δv value of 2.37 km/s does not include the budget for traveling from the surface of the Earth to LEO (over 9 km/s Δv), performing a translunar injection (over 4 km/s Δv), and landing softly on the lunar surface (over 2 km/s Δv)—although those preferring to lithobrake can save the last 2 km/s Δv! Nevertheless the "return" budget is more important, since the return voyage of a mining mission would contain orders of magnitude more mass (from the mined resources) than would be transported during the outbound phase (containing only the spacecraft and mining equipment).

isotope. For instance, to extract just 1 kg of ^{3}He from a high concentration (20 parts per *billion*) region would require processing 30,000 m^3 of regolith, or around two American football fields (including end zones) dug 3 m deep (or 12 Olympic-sized swimming pools 2 m in depth). (Note that concentrations of solar wind implanted volatiles decrease with depth, so there is a relatively shallow depth limit for the extraction of these materials from the lunar regolith.) Though there is roughly 2×10^6 km^2 surface area available in high ^{3}He concentration regions, with the potential for extracting roughly 2×10^8 kg of ^{3}He (145), extracting significant quantities of ^{3}He would be energy-intensive and highly destructive of the lunar surface.

Metals and Silicates

Iron, titanium, aluminum, and silicon are each abundant on the Moon. Iron, titanium, and water can be extracted from ilmenite (FeTiO$_3$) via hydrogen reduction (150). Silicon makes up roughly 20% of the mass of most lunar rocks, and aluminum accounts for 10% to 18% of the weight of regolith, at least in the highland regions (150–151).

Water

The lunar north and south poles are home to a number of permanently shadowed craters, i.e., impact basins where the crater rims permanently block sunlight. The north pole has around 13,000 km^2 of permanently shadowed area; while the south pole has nearly 18,000 km^2 area in permanent darkness (146). Altogether, then, the Moon has roughly 31,000 km^2 of permanently shadowed surface. These areas had long been thought to contain cold traps where water ice might exist. The presence of water ice was confirmed in 2009 by the Lunar Crater Observation and Sensing Satellite (LCROSS) together with the Lunar Reconnaissance Orbiter (LRO). When LCROSS deliberately impacted the floor of Cebaus crater in the south pole, LRO measured water in the ejecta at a concentration of 5.6 ± 2.9 parts per million (146). Though the depth of ice deposits is unknown, if all of the permanently shadowed area contains water at this concentration, then there is roughly 2.9×10^{12} kg water ice available in the first meter (146). This is about the same volume of water as contained in New Melones Lake (CA, USA) or in Trinity Lake (CA, USA), and would produce a sphere slightly less than 1.8 km in diameter.[8]

8. See Fig. 5.2 for a visualization of how this quantity of water compares with others, including the water available from near-Earth asteroids.

Peaks of Eternal Light

Another valuable resource located at the lunar poles are the peaks of eternal light, i.e., areas of the lunar surface (mostly along ridges and crater rims) that are nearly always exposed to sunlight. Though the only areas of 100% illumination are located in the lunar north pole, the south pole has a number of areas of ≥ 70% illumination conditions; many of these areas are along the rims of permanently shadowed craters. Thus, these peaks of eternal light are incredibly valuable lunar real estate—not only are they areas where one can collect solar energy without (or nearly without) interruption, but many are located very close to potential sources of water ice. But these areas are also incredibly rare, and all areas of ≥ 70% illumination together only account for a few hundred square meters of the lunar surface (mostly strips a few meters wide; see Fig. 5.1) (Elvis, Milligan, and Krolikowski 2016, 32).

Near-Earth Asteroid Resources

Other than the Moon the only other "nearby" resource base is the population of near-Earth objects (NEOs), primarily near-Earth asteroids (NEAs) as well as a small number of near-Earth comets (NECs). A NEO is generally defined as an object with a perihelion of < 1.3 AU.[9] To date, via both ground- and space-based observation platforms, 20,333 NEOs have been discovered, with 8,620 NEOs > 140 m in diameter, and 895 NEOs > 1 km in diameter.[10] Estimates for the total NEO population vary considerably. NEO size and classification is inferred from spectral analysis, color, and albedo measurement. Interpolating from these observations, distribution models provide estimates for the total NEO population, which is thought to contain between 910 and 990 NEOs > 1 km in diameter; 10,000 to 29,000 NEOs > 140 m in diameter; 10 million NEOs > 1 m in diameter; and 300 million to 500 million NEOs in total.[11] NEO resource inventories are estimated by combining size, distribution, and classification estimates with analyses of meteorite samples. According to one estimate, the NEO resource inventory includes 75×10^{15} kg silicates; 10×10^{15} kg ferrous metals; and 5×10^{15} kg water (Lewis 2015, 101).

---

9. The perihelion of an object in solar orbit refers to the lowest point in its orbit; one AU (astronomical unit) is approximately the average distance from the Sun to the Earth.

10. Discovery Statistics, Center For Near Earth Object Studies, Jet Propulsion Laboratory, available at: https://cneos.jpl.nasa.gov/stats/totals.html (accessed 8 June 2019).

11. These estimates are from Harris and D'Abramo (2015) and Anthony and Emami (2018).

North > 88

South < –88

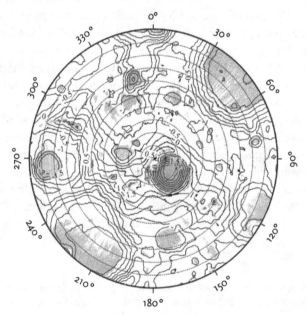

Figure 5.1 A visualization of both the permanently shadowed surface (light grey); areas of 60–70% illumination (medium grey dots); areas of 70–80% illumination (dark grey dots); and areas of ≥ 80% illumination (black dots) above the 88th lunar parallel north and below the 88th lunar parallel south. The distance from edge to edge is slightly greater than 121 km.

Credit Line: Reproduced from: H. Noda, H. Araki, S. Goossens et al., "Illumination Conditions at the Lunar Polar Regions by KAGUYA(SELENE) Laser Altimeter," *Geophysical Research Letters* 35 (2008): L24203. Copyright 2008 by the American Geophysical Union and John Wiley and Sons.

Figure 5.2 The relative spherical volumes of (from left to right) Lake Okeechobe (5.2 km³); NEO water (Sanchez and McInnes 2013) combined with Lunar polar water (Crawford 2015) (3.7 km³); Lunar polar water (2.9 km³); and NEO water (0.8 km³). The diameters of the spheres are 2.62 km, 1.92 km, 1.77 km, and 1.15 km, respectively. The Burj Khalifa (828 m) is shown for scale.
Credit Line: James S. J. Schwartz.

As mentioned earlier, it is necessary to factor for limited Δv budgets. Staggering resource totals such as those Lewis gives do not provide a viable assessment of the quantity of resources we are likely to access, for the simple reason that it would not be feasible to exploit the resources of every single NEO. Many of these objects are too energetically distant and would provide insufficient return on investment. A more realistic assessment of the available NEO resources must therefore account for restrictions due to limitations on Δv budgets. There are, however, a variety of methods for narrowing the NEO pool. Joan-Pau Sanchez and Colin McInnes (2013) provide information about the total NEO mass accessible at a return budget less than or equal to lunar escape velocity (2.37 km/s Δv). Martin Elvis (2014), in contrast, identifies the number of mining candidate asteroids at a different Δv budget (an *outbound* Δv of ≤ 4.5 km/s). Robert Jedicke et al. (2018) identify the number of mining candidate asteroids at yet a third Δv budget (an *inbound* Δv of ≤ 3 km/s). Lastly, Andrew Rivkin and Francesca DeMeo (2019) estimate the number of water-mining candidates with an *outbound* Δv of ≤ 8 km/s. Due to the incommensurability of the methodologies used in forming these estimates, I shall not attempt to draw direct comparisons between them but instead simply relay their results.

Sanchez and McInnes (2013)

According to Sanchez and McInnes, NEOs of diameter > 1 m altogether mass 5×10^{16} kg (2013, 447). They estimate that around 10^{14} kg of this material can be returned to cislunar space at a Δv of ≤ 2.37 km/s, with 8.5×10^9 kg returnable at a Δv of $\leq$ 100 m/s (447). With their estimates of NEA distribution and mass densities it is possible to construct estimates of the total resources available at these energy thresholds. Carbonaceous (C-type) asteroids are the most attractive water mining candidates. Given that these comprise 10% of the mass of the NEO population, and are approximately 8% water by mass (454), we can expect there to be approximately 8×10^{11} kg water from C-types accessible at $\leq$ 2.37 km/s Δv. This would correspond to a sphere of water about 1.15 km in diameter; see Fig. 5.2 for a visualization of this compared with Crawford's estimate of the water available in permanently shadowed lunar craters. A return Δv threshold of 100 m/s would yield 6,800,000 kg (6,800 m³) water from C-types—not quite enough to fill three 2 m deep Olympic-sized swimming pools.

Platinum-group metals (PGMs) are expected to be present at around 35 ppm in M-type asteroids, which comprise 5% of the NEO population and have an expected density of 5300 kg/m³ (454).[12] Thus, we can expect a total of 175,000,000 kg of PGM from M-types accessible at a Δv of $\leq$ 2.37 km/s, and a total of 14,875 kg PGM from those accessible at a Δv of $\leq$ 100 m/s. On average, one would have to process 29,000 kg of M-type asteroid material to produce 1 kg of PGM (corresponding to an M-type asteroid of about 2.2 m diameter); producing 1 ton of PGM would require the complete processing of an entire 22 m diameter M-type asteroid. However, M-types are estimated to be roughly 88% metal (mostly iron and nickel), and thus can provide considerable quantities of more common metals (454).

Elvis (2014)

In contrast with Sanchez and McInnes's mass-estimate approach, Elvis provides an estimate for the number of mining candidate asteroids. While his estimates do not provide the basis for a direct assessment of the *total* resources accessible from the NEO population, they do offer an account of how many retrieval missions might have to be carried out at specific Δv thresholds. He identifies an *outbound* Δv of 4.5 km/s (from LEO) as a benchmark, noting that at this threshold one can access 2.5% of the NEO population (2014, 20). Elvis estimates that around 4% from this population are the M-types which are suitable for PGM mining. Among these he expects only 50% to be rich enough in PGM to justify the costs of retrieval and processing. Further, he claims that a 100 m minimum diameter would be needed in order to produce enough PGM to provide an adequate return on investment (22). Starting from an estimated 20,000 NEOs of diameter $\geq$ 100m this only leaves on the order of

---

12. Note that at 35 ppm PGM, M-types compare favorably to terrestrial sources, with ores typically falling in the range of 5–15 ppm PGM (Zientek et al. 2017, N4).

Figure 5.3 An 18m diameter asteroid alongside both a sphere representing the volume of water (318 m³) contained within a C-type asteroid of this size, and a 1.8 m (6 ft.) human for comparison. The volume of the asteroid is comparable to the volume of a 2 m deep Olympic-sized swimming pool.

Credit Line: James S. J. Schwartz (water sphere); ESO/L. Jorda et al., P. Vernazza et al. (asteroid).

10 suitable PGM mining candidates! The figure for water mining of C-types is somewhat less austere. Working from the assumptions that C-types comprise 10% of the NEO population, with 31% being sufficiently rich in water, with 3% being accessible at a Δ*v* of 4.5 km/s, and with a minimum diameter of 18 m needed for sufficient returns, Elvis expects there to be on the order of 9,000 suitable water mining candidates. Using the mass density estimates provided by Sanchez and McInnes, an 18 m C-type asteroid would contain slightly less than 4,000,000 kg material, and at 8% water by mass it would contain nearly 320,000 kg water (320 m³); see Fig. 5.3 for a comparison of the respective sizes of such an asteroid and its water content.

Jedicke et al. (2018)

Jedicke et al. provide a third measure of the quantity of water accessible from the NEO population (2018). One feature which distinguishes their measure is that it is time-indexed. That is, they account for the fact that asteroids can only be accessed at minimal Δ*v* thresholds when their orbits and Earth's are properly aligned. Windows of opportunity for low energy transfers to candidate NEOs may only occur a handful of times a year, and so Jedicke et al.'s estimate of the number of mining candidates is considerably lower than Elvis's estimate, at least on the decade-scale discussed in their article. They assume a return Δ*v* threshold

Figure 5.4 The relative spherical volumes (from left to right) of ten years (8,400 m³); one year (840 m³); and one month (70 m³) of NEO water mining at expected average rates (Jedicke et al. 2018), next to a 1.8 m (6 ft.) person for comparison. The diameters of the spheres are 25.2 m, 11.7 m, and 5.1 m, respectively.
Credit Line: James S. J. Schwartz.

of 3 km/s (29), and that C-types make up 16% of the population (by number) accessible at this threshold (36). Restricting their focus to candidates between 5 m and 10 m in diameter, they identify 1,000 water mining candidates (37). Given the timing restrictions due to the varying orbital profiles of NEOs, they estimate that, on average, five such asteroids could be returned per month, yielding an average of 70 tons of water per month (39). At this rate (which would vary in practice) it would take a five-month water harvest to fill a volume equal to the *Saturn V*'s third stage tanks (326,000 L); a 19-month harvest to fill a volume equal to the *Saturn V*'s second stage tanks (1,287,000 L); a 29-month harvest to fill a volume equal to the *Saturn V*'s first stage tanks (1,974,000 L); and a three-year harvest to fill a 2 m deep Olympic-sized swimming pool (see also Fig. 5.4).

Rivkin and DeMeo (2019)

An even more recent estimate (Rivkin and DeMeo 2019), which focuses on C-type NEOs with an outbound Δv of 8 km/s, anticipates that there exist 19 ± 9 mining candidates of diameter > 1 km, and 1800 ± 900 mining candidates of diameter > 100 m. Rivkin and DeMeo assume a water content of about 7% and a mass density of 1300 kg/m³ for C-types. This corresponds to a total of about $10 \pm 4.7 \times 10^{11}$ kg of water, or a volume of 1 ± 0.47 km³, and would produce a sphere of water ranging from 1 to 1.4 km in diameter (with most of this water bound up in the 1 km size objects).

The Significance of Scarcity

Resources in near-Earth space, then, are in certain cases very limited. Although the total quantity of water and PGMs potentially available from the Moon and NEOs is considerable, nevertheless *accessible* quantities are much smaller. While not the most attractive location from a Δv perspective, the Moon does represent something of a fixed quantity: It stays roughly the same distance from Earth, which fixes both the transit time between the Earth and Moon (about three days each way) as well as the energy required for transporting materials from the lunar surface to Earth orbit. Thus, the Moon is likely to remain a sought-after resource base irrespective of the status of asteroid mining.

NEOs, meanwhile, promise a greater variety of resources at varying Δv budgets. However, missions to NEOs generally will require on the order of years to decades for planning and implementation. During any given year only a small number of transfer windows will open to low-Δv candidates. Moreover, one-way transit times are likely to exceed six months (Jedicke et al. 2018). Thus, we are unlikely to benefit from a large influx of commercially viable materials from the NEO population. Instead, these resources would only arrive in a slow, staggered trickle. And as the first agents to mine NEOs are likely to target objects with the smallest Δv budgets and with the shortest time horizons, each subsequent generation of NEO resource retrieval will be increasingly expensive and increasingly time consuming. Those late to the NEO mining game may find few candidates that they can exploit profitably and quickly enough, validating Linda Billings's worry that "those with the means to get to the 'store' of space first get all the goods" while those "who get there late may get nothing" (2006, 252).

It is also worth noting that lunar and asteroid resources are for all practical purposes non-renewable. NEOs have a half-life of 10 *million* years (Jedicke et al. 2018, 32), and thus, once depleted, "it will take millions of years of orbital evolution until the population of asteroids in near-Earth space is replenished" (Sanchez and McInnes 2011, 163). The concentrations of solar-implanted volatiles in the lunar regolith are the result of *billions* of years of solar activity, while the presence of water ice in permanently shadowed craters is the outcome of *billions* of years of comet impact events. The present concentrations of these resources, once depleted, will not be replenished prior to the expansion of the sun into a red giant.[13]

The upshot is that the resources of NEOs and the Moon are far from limitless or inexhaustible. This vitiates premise (2) of the schematic argument

---

13. NEOs likely start their lives as Main Belt objects (perturbed into orbits with perihelia below 1.3 AU). So even though NEO resources aren't replenished terribly quickly, it is always possible to head directly to "the source" in the Main Belt to support longer-term asteroid mining.

for space mining and for implementing a regulatory regime friendly to space mining. Because there is a relatively small upper bound on some of the most valuable space resources, and because these resources would only arrive slowly, irregularly, and in very small quantities, their exploitation would not help much to, e.g., mitigate terrestrial resource depletion, protect Earth's ecosystems by moving polluting activities to space, or improve the material well-being of majority of humanity. For these reasons we ought to reject as naïve attitudes like those expressed by Cooper when he claims:

> Human beings cannot remotely conceive of the depletion of all our solar system's resources. Even given the lessons of our recent past there seems enough for all, if the effort is made to access the resources. (2003, 114)

The problem is that, for the moment, most of the Solar System's resources are too distant. We can conceive quite easily the depletion of *accessible* space resources. Unless the use of exploited resources is managed carefully, the rest of the Solar System will remain hopelessly out of reach.

SPACE RESOURCES: BENEFITS FOR HUMANITY?

The relative scarcity of resources like water and PGMs undermines an important pillar of arguments in support of commercial space mining. We should be very skeptical of claims about the potential for space resource exploitation to solve all terrestrial resource and energy problems. But that is not all we should be wary of. Another promise often made by space-mining advocates is that the growth of our resource base through space mining will increase overall wealth in ways that will significantly benefit all of humanity. Given the limited pool of accessible space resources, the rate of any such economic growth is likely to be miniscule. Still, it is difficult to believe, were we to grant to space mining firms the legal protections they desire, that the result would be anything other than an exacerbation of existing inequalities, both at the national and international levels. As Alan Marshall argues, "the development of even more resources is not likely to provide for the necessities of most of the world's people" (1995, 42). Instead, he claims, space resources would "contribute to the consumptive wants of the wealthy, not to the needs of the populous poor" (42). Tony Milligan is similarly skeptical about the willingness on the part of space mining firms to contribute to genuine societal progress:

> [I]n the absence of a radical alteration in patterns of human behaviour, a good deal of energy from ^{3}He mining is unlikely to go towards a great life-enhancing project. It is likely to be used for comparatively trivial purposes such as advertising, waste and the enhancement of prestige. (2013, 5)

As we have witnessed (especially during the last few decades), the free market is an unreliable tool for improving the well-being of the least fortunate members of society. Though economies have grown, so too have inequalities, with wealth and political power becoming increasingly more concentrated in the hands of the fortunate, powerful few.[14] Were space mining advocates truly interested in helping the least fortunate among the human population, then they would favor a more communitarian approach to space resource exploitation, for instance as mandated by the ill-fated Moon Agreement.

Of course, when presented with redistributive frameworks, space mining advocates often recoil in horror, double down on the alleged benefits of a free market system, and reaffirm the limitless nature of the space environment. For instance, Lewis Solomon notes that developing states

> will likely fight changes designed to promote private sector space activities. However, space resources may be nearly limitless, thereby negating the need for benefit sharing, such as done in the Moon Treaty. Even if developing nations take decades to be able to tap space resources, these resources will likely remain plentiful. (2008, 112)

Remarks such as these are naïve at best. It is by no means guaranteed that developing nations will *ever* be able to afford to support space resource exploitation missions, especially in the case of states that already find it difficult to feed and care for the majority of their citizens. But even if, decades down the line, a developing state were to build the necessary capacities for exploiting space resources, they may well find that other states' earlier mining initiatives leave them with no mining candidates within an affordable Δv budget. As Milligan notes, "[i]t is no good depleting all that can be reached and then pointing to what remains at an impossible distance" (2015b, 127).[15] It is also possible that advocates are sometimes being disingenuous when proselytizing about benefits to the least fortunate, as evidenced by Pop's entreaty that "have not nations," rather than "freeloading on the efforts of the older spacefarers," should instead

---

14. Natural resource dependence has been identified as a potential contributor to inequality and economic stagnation. As ElGindi (2017) argues, the impact on income inequality varies on the basis of, *inter alia*, the type of resource under consideration, with mineral-rich countries exhibiting higher levels of income inequality. Meanwhile, Sun et al. (2018) suggest that investment in public education can help to mitigate economic stagnation within natural resource dependent communities.

15. *Cf.* Saletta and Orrman-Rossiter (2018, 2): "[T]here is a 'Silicon Valley' venture capitalist marketing spin in the discourse surrounding the prospect of commercial resource exploitation in space, characterized by appeals to the mythos of the Wild West, gold rushes and with not infrequent echoes of Manifest Destiny. While companies and entrepreneurs justifiably intend to enrich their investors, claims that this will in turn enrich humanity more generally sound suspiciously like trickle-down economics."

pool their financial resources into a common space agency or into regional ones, and proceed at exploiting the riches of outer space themselves. Indeed, they need not build the infrastructure themselves if they can buy commercial space services on the global market. (2012, 562–563)

So, developing nations need not fear missing out on space resources because they can always pay developed, space-capable states to mine resources for them (no doubt at a premium)! It is the height of absurdity to insist that such a system would attenuate rather than exacerbate inequalities.

A minimally regulated space mining industry would not, of its own volition, prioritize the extraction of those resources most beneficial to the majority of humanity (except accidentally if this happened to coincide with market forces). Nor would any profits derived from such exploitation benefit anyone except the already fortunate, except possibly through marginal increases in the tax revenue base (assuming the industry does not successfully lobby for tax exemptions). For space mining to be *truly progressive*, that is, for it to *genuinely benefit humanity as a whole*, as opposed to a small number of already wealthy individuals, it must be redistributive.[16] Unfortunately such redistribution is exactly the opposite of what the space mining industry appears to want. Worse still, there are not many space resources that even have the potential to benefit humans that remain on Earth. While terrestrial supplies of nickel and PGM may run so thin as to make asteroid sources competitive in terrestrial markets, it would be absurd to say that humanity as a whole would benefit from, e.g., being supplied with water or aluminum from space. For resources like water or aluminum to become cheaper to buy from extraterrestrial sources would imply that the situation on Earth had deteriorated beyond a reasonable point. That is, if it is cheaper for terrestrial humans to purchase their water from space, then they are already doomed. The most likely beneficiaries of resources derived from the Moon and NEOs would be space-based consumers, e.g., humans living in orbital or lunar settlements. Indeed, it is only once consumers are already in space that the exploitation of space resources becomes economical. For many reasons, then, we should not hold out hope that the exploitation of space resources will make much or any difference to the lives of the overwhelming majority of terrestrial humans. It follows that the alleged socioeconomic benefits of space mining do not

16. As Armstrong (2013) argues, it is not clear exactly what kind of redistribution would genuinely benefit humanity *sensu* promoting social justice. On Armstrong's analysis, a just distribution of goods addresses "inequalities across the range of advantages and disadvantages relevant to egalitarian justice," and these are not exhausted by inequal access to natural resources and benefits derived from natural resources (333). Thus, an overall just distribution of goods might (or might not) include a great deal of inequality when it comes to natural resources, and the same might hold with respect to access to resources derived from space mining.

provide a compelling basis for encouraging space mining through industry-friendly legislation.

IMPLICATIONS FOR FUTURE POLICY

Law and policy, though ideally motivated ethically, in practice involve compromises between interested parties and favor predictability, even if the result is flawed in many ways. In this respect having a legal system, even a highly flawed one, might be better than having no legal system at all.[17] This point applies to the regulation of space mining. It would be naïve to pretend that the above arguments would, in practice, compel a moratorium on space mining, and nor would I argue that space mining ought to be permanently shelved. Sooner or later space mining will probably take place, and it would be best for all concerned if it takes place under a regulatory framework, even if that framework leaves much to be desired. But even if an ethically ideal framework is beyond reach, there is still an opportunity for the promotion of a system that is closer to the ideal rather than further from it. As Christopher Newman and Mark Williamson note, there is time for "technical and legal professionals to work in harmony to provide optimal solutions" that ensure the sustainability of space resource exploitation and use (2018, 34). To such ends, I want to close this chapter by describing how the unjustly neglected issue of space resource scarcity might be incorporated in future space policy and law discussions.

Various existing regimes have been proposed as models for the regulation of space mining: The Outer Space Treaty; the Moon Agreement; the International Telecommunication Union's (ITU) Orbital Allocation Regulations; the Antarctic Treaty System; the U.N. Convention on the Law of the Sea; etc. Providing an assessment of each of these regimes would take us too far afield, and in any case, what I would like to focus on is a feature common to each of these systems, and that is an emphasis on *fair access* to locations and/or resources. The ITU Orbital Allocation Regulations provide a good example of what I am after. These regulations manage allocations in various Earth orbits, including geostationary orbit (GEO) where the orbital period matches Earth's rotation. An orbital allocation includes not only an orbital position, but also a radio frequency allotment (to prevent communications interference between neighboring satellites). Allocations in GEO are especially valuable resources because a satellite placed in GEO remains "in place" from the perspective of someone on the surface of the Earth. They are also limited resources, since only a finite number of satellites can operate in GEO

17. Thanks to Frans von der Dunk for discussion on this issue.

at any given time. The finite, limited nature of GEO allocations has long been acknowledged by the ITU, and according to Article 44 of the ITU Constitution,

> radio frequencies and any associated orbits, including the geostationary-satellite orbit, are limited natural resources and . . . must be used rationally, efficiently and economically, in conformity with the provisions of the Radio Regulations, so that countries or groups of countries may have equitable access to those orbits and frequencies, taking into account the special needs of the developing countries and the geographical situation of particular countries.

In practice, the ITU's policy of ensuring "equitable access" to GEO allocations involves both an *a priori* and an *a posteriori* plan.[18] According to the *a priori* plan, the ITU reserves a GEO allocation for every state regardless of its space capabilities. All other allocations belong to the *a posteriori* plan, which is first-come, first served. Whether this arrangement provides *genuinely fair* access to GEO can and perhaps should be disputed (Schwartz 2015c); my point is simply that the policy *intends* to provide such access. Beyond this, the ITU cares naught for what *purposes* operators use their GEO allocations. Were a wealthy individual interested in placing an electric car in GEO (and assuming they acquired a launch license from their launching state), the ITU could not, and would not, cite the frivolousness of this proposal as a reason to deny the application for a GEO allocation. Thus, the ITU regulations solely manage *access* to orbital allocations; they in no way regulate the *use* of these resources.

The Moon Agreement provides a second example of a preoccupation with *access*. Article VI reaffirms from Article I of the Outer Space Treaty that "there shall be freedom of scientific investigation on the moon." Articles VIII and IX hold that states are free to land personnel and equipment on the Moon but that in the process states are obligated not to impede other states' "free access" to the Moon. Paragraph 4 of the infamous Article XI indicates that states "have the right to exploration and use of the moon without discrimination of any kind, on the basis of equality." Moreover the goal of the international lunar resource exploitation regime, as mandated in Article XI, is tasked with ensuring the "equitable sharing by all States Parties in the benefits derived from those resources" (Par. 7). Thus, not only does the Moon Agreement strive to ensure *fair access* to the Moon itself, but also *fair access* to the *benefits derived* from the exploitation of lunar resources.

Laudable as these regimes are for their intention to ensure fair access to space (and in the case of the Moon Agreement, to the benefits derived from lunar mining), they each suffer from the same damning lacuna. To use a

18. See Jakhu (2017) for a more detailed overview.

somewhat melodramatic example, suppose that an incredibly wealthy space entrepreneur wanted to sculpt a 1-km tall likeness of their person out of lunar water ice. Suppose further that under a Moon Agreement-like system, this person was willing to pay the international regime fully in order to contract the necessary exploitation, and that this massive monetary transfer would be distributed equitably to all states. Such a distribution would, then, satisfy the requirement to grant states fair access to the *benefits derived from* the exploitation of this water ice. But in this scenario something has gone terribly, horribly wrong in a way the Moon Agreement is powerless to prevent: A *significant* percentage of the incredibly rare lunar water ice deposits—which could have been used to produce drinking water, air, and rocket propellant, have instead been pilfered for an entirely frivolous purpose—the vanity of a single, wealthy individual. No reasonable regulatory system would permit such a deliberate waste of an exceptionally limited, life-giving resource. For a scarce space resource like lunar water ice, there is little benefit to the regulation only of *fair access* to resources, since there may not be enough in the first place for everyone to acquire a useful share. Nor is it enough to regulate access to the *benefits derived* from resource exploitation, either because the benefits would be worth too little if distributed widely, or because the resources themselves would go to waste. What is missing from existing regulatory proposals, then, and what is sorely needed, is a framework for regulating the ways that space resources are *used*. In the case of lunar polar water ice, for instance, we cannot afford to use this limited resource in support of anything other than the most ethically defensible projects. While the example of a 1-km tall ice sculpture is fanciful, something more than mere hope in market forces is required for avoiding more realistic but similarly unproductive uses of scarce space resources.

An ethically informed regulatory system must recognize that not all *uses* of the space environment or its resources are of equal value, and that certain uses ought to be encouraged by means of enabling legislation, whereas other uses ought to be discouraged by restrictive or prohibitive legislation. The ethical permissibility of an instance of space resource exploitation is a function of two factors related to use. First concerns the use of the *environment* in which exploitation might take place, as other stakeholders may have compelling interests in the *site or location* of mining activity. An asteroid, for instance, is not merely of interest to mineral prospectors, but also to space scientists—not only can asteroids be mined for resources; they are also legitimate objects of scientific curiosity, as Shepard (2015) relates in great detail. Scientific interests ought not to always lose out to commercial interests. Second is whether the *exploited resources* will be *used or consumed* in a manner that is ethically defensible. Lunar polar water ice, as we have already seen, can be used by many different stakeholders for many different purposes, not all of which should be permitted.

This, of course, raises the question as to which uses of space and space resources we ought to encourage and which uses we ought to discourage. As I have argued in this chapter, the moral case for the exploitation of space resources for, e.g., ameliorating terrestrial environmental crises, improving the overall condition of humanity, etc., is not strong. Although we are obligated to take measures to solve terrestrial environmental crises and to better provide for humanity *in toto*, nevertheless lunar and NEO mining are not effective means for satisfying these obligations, at least over timescales measured in decades rather than centuries or millennia. For the moment, our efforts would be better spent on terrestrial solutions to terrestrial problems. I am similarly skeptical about the urgency or effectiveness of using space and its resources for human space settlement. Summarizing very briefly one of the claims I defend in the next chapter: While we do have an obligation to take measures to ensure the long-term survival of the human race, efforts to settle space will not effectively help us to satisfy such an obligation over the near future. Only a very small amount of good would come from the near-term use of space and its resources for profit and for settlement. The use of space and its resources for, e.g., destination resorts for the wealthy, would produce even less overall good!

Meanwhile, as I argued in Chapters 2 and 3, there is considerable good that can be realized from the scientific investigation of the space environment together with the increases in scientific knowledge and understanding that such investigation brings. *Ceteris paribus*, if we face a decision between preserving a space environment for scientific study, and using the same environment as a mining site, we ought for the time being to opt for the former. Similarly, other things being equal, if we face a choice between using space resources to support scientific exploration missions and using space resources to supply mining prospect missions or space hotels, we again ought for the time being to opt for the former. An *ethically grounded* regulatory system, then, must include mechanisms which prioritize scientific uses of space and space resources over non-scientific proposals.

The potential for conflict between scientific and other uses of space is real. Already there is competition over limited orbital allocations in GEO and other Earth orbits, and at some point in the not so distant future we will have to decide between launching satellites for the purpose of scientific observation and launching satellites for the purposes of commercial or military services. Further afield, if commercial spaceflight operations increase, we are likely to see conflicts of use on the Moon and on asteroids. The same sobering details of orbital mechanics and time horizons that severely limit the number of NEO mining candidates also severely limit the number of candidates for scientific missions to NEOs (Krolikowski and Elvis 2019). A peak of eternal light might be entirely co-opted by a single solar power station, or a single scientific experiment (Elvis, Milligan, and Krolikowski 2016; Wingo 2016). We require a regulatory system that is capable of making rational, ethical decisions in such cases.

Scientific and other interests are not incompatible in all cases. For instance, several scholars have noted that a commercial asteroid mining industry might help to facilitate the scientific study of asteroids (along with other knock-on effects such as the protection of Earth from asteroid collisions). Philip Metzger explains:

> At the current rate of scientific investigation, we could visit all [asteroids over 10 m in diameter] in about 10 billion years, except the sun will not last that long. Mining will cause us to see far more than we would otherwise. The same argument can be made for lunar mining, which may be largely underground to escape the radiation problem, and therefore will reach materials that would not otherwise be accessible for science, but which have immeasurable scientific value. (2016, 81)

In addition to providing *access* to space for science, space resource exploitation would also provide science with resources which could help sponsor more ambitious exploration missions by:

> making rocket propellant from asteroidal or lunar volatiles, building radiation shields and landing pads from regolith, [and] providing breathing oxygen and water for outposts to make metal for 3D printing spare parts. Those activities will make space missions more effective because they reduce the amount of mass that must be launched from Earth while providing more flexibility during missions to respond to problems. (80)

Other scientifically beneficial items that could be manufactured out of space resources include large in-orbit reflectors for optical and radio astronomy (McInnes 2016), as well as many other scientific instruments, supplies, and outposts (Crawford 2016; Elvis 2016). But in the absence of legislation supportive of science, we have no guarantee that mining companies would give scientists access to data and samples. Nor do we have any guarantee, again in the absence of legislation supportive of science, that any benefits derived from exploited space resources would wind up in the hands of scientists. Though *potentially* of great benefit to science, it is very much open whether commercial exploitation would benefit science in practice, since the private sector would be likely to regard asteroid locations, trajectories, and resource inventories as proprietary information. Promises to the contrary should not be taken on faith.

It is also worth reminding the reader of an important point from the previous chapter, *viz.*, that scientific exploration is most effective in *pristine* environments. Secondhand access to lunar or asteroid materials, perhaps via occasional donations from mining corporations, may be better than nothing, as far as science is concerned. But such samples might provide insufficient geologic context, and would thereby limit the strength of conclusions scientists could draw from their analyses of the samples. This point is put nicely by Ehrenfreund, Race, and Labdon:

> [R]egions on the Moon provide a window to the origin of our solar system and the Earth. Outstanding scientific findings about our origins are expected from further investigating the cratering record, lunar soils, icy polar regions, and the exosphere, to name a few. Despite the Moon's seemingly dead nature, physical disruption of the surface by frequent landings and launches may compromise the potential to gather scientific information. In addition to disruptions from landings and lift-offs, activities like EVAs, site destruction through human activities, dust raising, atmosphere contamination, vibration, radio contamination, and use of nuclear power sources can adversely affect scientific research. (2013, 67)

Similar considerations apply to asteroids, as Krolikowsi and Elvis explain:

> Asteroids are most valuable to science in an unaltered and uncontaminated state. Many research goals can be pursued to the fullest only if the asteroid under study has not been previously modified for another purpose. For these reasons, scientific efforts to characterize or modify asteroids are likely to be delicate, localized endeavors designed to minimize their impact on the surrounding material and environment. (2019, 12)

That the scientific study of the space environment is relatively benign raises another salient point. Resource exploitation, which generally is entirely destructive of an environment, can without significant loss be undertaken preceding scientific study, whereas scientific study can only be conducted most effectively prior to exploitation. That is, the cost to science of exploitation is much greater than the cost to exploitation of science. Moreover, prior scientific study can be seen as an enabler of exploitation, since it would save commercial miners from some of the costs and troubles associated with resource prospecting and exploration.

Various factors, then, converge on highlighting scientific interests as most worthy of prioritization when it comes to the regulation of the use of space and space resources. Not only does science promise greater ethical rewards; its interests are also more vulnerable. As far as actual policy proposals are concerned, I have two suggestions to make.[19] My first suggestion relates

---

19. I proposed earlier versions of these suggestions in Schwartz (2016b) and Schwartz and Milligan (2017).

to decisions about which locations might be permitted as mining sites. As identified in the previous chapter, science has yet to disclaim interest in any particular aspect of the space environment. The presumption, then, should be one that favors science. We should assume that an environment or entity is scientifically valuable until proven otherwise. Resource extraction should only be permitted if the scientific community has decided to disclaim interest in whichever environment/s would be disturbed or destroyed in the course of the extraction.

A possible international policy mechanism here would be for the U.N. to convene a scientific panel (ideally with the involvement of an organization like COSPAR) tasked with evaluating the scientific value of locations identified in mining proposals. Such a body should be granted veto power over mining proposals, and should be made as free from external influence and corruption as possible. Still, since the course of science is unpredictable, a proposal approved by scientists in 2050 may later be deemed to have resulted in the destruction of a unique opportunity for scientific study. Such risks are unavoidable, but to minimize them, approved proposals should be subject to ongoing, independent scientific oversight. This way, in the course of mining, should anything of unforeseen scientific interest be discovered (e.g., evidence of life, anomalous mineral contents), timely requests could be made to halt exploitation and hand the materials or location over to the scientific community. A tax or insurance fund could be mandated that would provide remuneration to miners who in these situations are forced to revoke their mining rights. Details aside, the basic principle around which any legislation ought to be crafted is that mining should never be permitted in places of high scientific value, even if this value is not apparent until after mining operations have commenced.

My second suggestion relates to the use of extracted resources. Given the scarcity of most accessible space resources, restrictions must be placed on how these resources are consumed. Consumption in support of worthwhile scientific missions ought to be prioritized over consumption in support of human settlement, and each of these uses should be prioritized over consumption for the sake of space tourism. A number of mechanisms are possible here. For instance, a Moon Agreement-like system might involve an international regime as the initial purchaser of all exploited material, which then makes various amounts of resources available depending on the nature of the consumer. Scientific consumers might then be given greater or cheaper access to water than, e.g., space resort developers. Another possibility is a tax on exploitation revenues that contributes to a science fund, the proceeds of which either support science missions, or support the acquisition of resources for science. In lieu of a monetary transfer, exploiters might instead contribute a portion of the resources they have extracted, which the fund could then either sell on the market, or utilize directly in support of science missions.

I would not be surprised if the particular policy ideas just mentioned are for various reasons politically infeasible, but their feasibility is beside the ethical point, which is that there is a genuine need for future space resource policy to take seriously the scarcity of accessible space resources, and to do so in a way that identifies science as the primary stakeholder when it comes to the use of space and its resources. But even if one rejects that science ought to play such a privileged role in space law and policy, still, one would have to agree that the scarcity of accessible space resources calls for the regulation not simply of *access* to space but also of the *use of* space and its resources. *The incredibly vast array of resources available at higher Δv budgets (the rest of the NEO population; the Main Belt asteroids; the planets and their satellites; Kuiper Belt objects; etc.) will become permanently inaccessible to us if we do not reserve enough of our early spoils for the expansion of spaceflight capabilities.* The issue of sustainable space resource consumption must be brought to the fore of all future discussions about space resource exploitation.

CHAPTER 6

The Need for Forbearance in Space Settlement

An important lesson from the previous chapter is that even comparatively strong duties can be overridden. By the contrapositive of "ought
implies can," if we lack any effective means for attaining something, we have
no corresponding obligation.[1] Moreover, higher-level duties, such as the duty
to improve the overall human condition, can often be satisfied in multiple—
possibly harmonious, possibly exclusive—ways. A duty to improve the overall
human condition does not establish an obligation to satisfy this duty in any
specific way, unless some particular action is (practically) *necessary* for the satisfaction of the overall duty. We ought also to acknowledge the possibility of
conflicts of duty. If we face a choice between satisfying a comparatively weak
duty (for which we have effective, affordable means) and satisfying a comparatively strong duty (for which we have no effective, affordable means),
the outcome of our moral calculus may favor the former over the latter. This
is especially likely in cases where the conflict is transient and asymmetric—
where satisfying the stronger duty first would preclude satisfying the weaker
duty, but where satisfying the weaker duty first would not preclude satisfying
the stronger duty over the long-term. In the previous chapter I argued that
considerations such as these militate against an urgent or overriding duty to
exploit space resources. As I shall argue in this chapter, similar considerations
vitiate an urgent duty to settle space.

What I shall focus on below is the argument that we have an obligation
to settle space because doing so would improve the chances of the long-term

---

1. Of course, in some cases we might judge that, all things considered, we have an
obligation to devise effective means.

survival of the human species. Following Tony Milligan (2011), I will call this the "duty to extend human life." This, I take it, is the strongest possible defense available to settlement advocates, since the motivating obligation appears to be one of the strongest obligations there is. In The Case for Space Settlement I motivate two related arguments for space settlement. The *basic argument* for space settlement supports the conclusion that we have an obligation to extend human life via space settlement. The *urgent argument* supports the conclusion that space settlement must be accomplished as soon as possible. I explore various criticisms of the *basic argument* in Against the *Basic Argument*, including: two objections to the underlying obligation to extend human life; an objection that pursuing space settlement would conflict with a duty to protect terrestrial ecosystems; and an objection that space settlement would not be an effective means for satisfying the duty to extend human life. I show that neither argument succeeds in refuting the *basic argument*, but that the effectiveness objection refutes the *urgent argument*. While we have a duty to extend human life via space settlement, it is not a duty that requires very much of us in the present day.

I provide another criticism of the *urgent argument* in Against the *Urgent Argument*, where I argue that science activities ought to be prioritized over settlement activities when these two endeavors conflict. This objection is structurally similar to my objection to space resource exploitation from the previous chapter: Space settlement has the potential to significantly diminish the scientific value of any of the environments in which it is attempted. Nevertheless since the *basic argument* survives this criticism, I do not deny that, over the long term, space settlement is genuinely necessary for satisfying our duty to extend human life. However, as I explore in Human Societies in Space, the pursuit of space settlement raises ethical concerns well beyond those stemming from conflicts with science. The extremely hostile environments found in space will place great demands on settlers, who may be forced, on pain of settlement collapse, to accept deplorable living conditions with few personal liberties. Settlement advocates often dismiss such sacrifices as simply the cost of long-term species survival. I shall argue for an alternative perspective, according to which space settlement should not be undertaken without a reasonable guarantee that settlers and their descendants will not be coerced unjustifiably in their major life decisions and decisions involving bodily integrity. Otherwise space settlement would necessitate unacceptable exploitation of settlers.

THE CASE FOR SPACE SETTLEMENT

Nearly every rationale that has been offered in support of space exploration generally also has been offered in support of space settlement: that establishing settlements in space will inspire students to pursue STEM degrees; that space settlement will reopen the frontier; that it will spur social and technological innovation; that it will help preserve terrestrial ecosystems; and so on. Since I have already dealt with these kinds of justifications in Chapter 1 (and to a lesser extent in Chapter 5), I will not pause to rehearse my criticisms here. However, this leaves one important rationale untouched, viz., that we have an obligation to settle space because this is necessary for satisfying our duty to extend human life. This rationale is widely promulgated—to the point where it is endorsed on virtually every occasion in which space settlement is discussed.[2] Thus, what I intend to capture here is not the exact position of any particular individual, but instead the general position space settlement advocates have tended to adopt.

The case for space settlement often begins by noting the relative fragility of terrestrial human existence. Possible species-terminating homespun threats include: nuclear war; global pandemics; global warming; and resource depletion. But we also face extraterrestrial threats: asteroid collisions; solar flares; the steady increase in the Sun's energy output; supernovae of nearby stars; etc. In the face of these threats, our survival chances would be improved, possibly dramatically, by the establishment of permanent, self-sustaining extraterrestrial settlements. These would include orbital habitats, lunar and Martian settlements, as well as asteroid-, satellite-, and other planetary-based settlements.[3] As most of, if not all of, these locations contain all of the basic elements required for human survival, it would be possible in principle to settle in them (though in many cases this would be very challenging, technically and socially). The threat of human extinction is thereby predictable and remediable. It follows that our duty to extend human life obligates our pursuit of permanent, self-sustaining space settlements.

The *basic argument* for space settlement, then, runs roughly as follows:

1. We have a duty to extend human life where possible.
2. We face a variety of threats to human existence.
3. It is possible for us to evade these threats by pursuing space settlement; moreover, certain threats can be avoided only via space settlement.

———

4. Therefore, we have an obligation to pursue space settlement.

There are those who would go even further than this. For instance, Donald Barker (2015) sees great urgency in the duty identified in (4). He claims that "[d]elaying the migration of our species off Earth only serves to heighten the

2. For a brief sampling, see Kondo et al. (2003); Zubrin (2009); Schulze-Makuch and Davies (2013); Gilster (2013); Munévar (2014); Barker (2015); and Musk (2016).
3. Though Earth's habitable lifespan ranges in the millions of years, life in the Solar System may only have one or two billion years due to the Sun's eventual expansion into a red giant. Thus *interstellar* settlement may ultimately be necessary, though I shall not focus much on it here—but see Schwartz (2018b).

probability of catastrophe and extinction" (51) and thus that space settlement "needs to be implemented as soon as any culture becomes capable of undertaking if it desires to survive" (56). The idea here is that any significant delay of our attempt to settle space would be tantamount to negligence, since it would avoidably increase the risk of human extinction. Thus we might extend the basic argument in the following way:

5. Adequately evading threats to human existence requires pursuing space settlement as soon as possible.

    _____

6. Therefore, we have an obligation to pursue space settlement as soon as possible.

Call the argument from (1)–(3) to (4) the *basic argument* for space settlement; and call the extended argument from (1)–(5) to (6) the *urgent argument* for space settlement. The question arises as to which of these arguments we should endorse. Since the conclusion of the *urgent argument* implies the conclusion of the *basic argument*, one could not consistently endorse the former while rejecting the latter. However, it is possible to accept the *basic argument* while rejecting the *urgent argument*, for instance, by arguing that space settlement, while necessary over the long-term (say, at some point over the next 1,000 years), is not obligatory over the near future (a period I take to correspond to roughly the next two centuries). Meanwhile, one could reject both arguments by undermining the existence of an obligation to pursue space settlement. As I shall argue in Against the *Basic Argument*, several common methods for rejecting the *basic argument* are ultimately unsuccessful—thus we have an in-principle obligation to pursue space settlement. Nevertheless, as I argue in both Against the Basic Argument and *Against the Urgent Argument*, we should go no further than this. Although we ought to accept the *basic argument* for space settlement, we ought to reject the *urgent argument*. Space settlement, though ultimately necessary, will for a considerable time remain an ineffective means for extending human life when compared to alternative strategies for evading most threats to human survival. Furthermore, a policy of urgent space settlement would conflict with other duties, and in particular, it would conflict with the duty to conduct effective scientific exploration of the Solar System, as human habitation of the space environment would disrupt the largely pristine and unexplored space environment.

AGAINST THE *BASIC ARGUMENT*

I will consider three strategies for undermining the *basic argument*. One route is to argue against premise (1). I shall explore two such arguments below. The first argument conceives of the duty to extend human life as a duty to future generations, maintaining that such duties are conceptually confused. The second argument denies that anything of value would be achieved by extending human life. A second strategy is a *reductio* which claims that the *basic argument* would have the unacceptable consequence of inuring humans to terrestrial environmental problems, thus hastening the degradation of Earth's ecosystems. A third strategy undermines premise (3) by arguing that space settlement would not be a particularly effective means for extending human life. Below I offer reasons for rejecting both the first and second strategies. While I provide a qualified endorsement of the third strategy, I do so in a way that leaves the conclusion of the *basic argument* intact.

Duties to Future Generations

A duty to extend human life is not implausibly understood as a duty involving future persons. However, according to an influential line of reasoning, duties to future persons are incoherent. If that is right, then the same fate befalls the obligation to extend human life. What I am referring to here is the "Non-Identity Problem."[4] The Non-Identity Problem begins with the observation that human persons have their genetic makeups essentially. For instance, had the gametes that produced the zygote which resulted in actress Claudia Black's gestation and birth never paired, she would never have existed. And these gametes having paired to form this zygote was a highly contingent affair. Had her parents not attempted to conceive precisely when they did, other gametes would have paired, and her parents would have conceived some other person (much to the detriment of the TV series *Farscape*!).

Consider, then, a proposal that would radically increase our consumption of non-renewable resources. This would affect who conceives and when. Several generations down the line, it is unlikely that any specific person would be common both to this future and a future where the status quo is maintained. Suppose we choose the policy of increased consumption, with the result that after 200 years the Earth is a desolate wasteland barely capable of supporting human life. Could a person living in this future claim a wrong against our generation, to the effect that our policy of increased consumption was harmful to them? It seems not, because had we maintained the status quo, this person would never have come into existence, and thus would not (and could not) have been harmed or wronged. Generalizing, a policy of increased consumption does not make *any* person living 200 years from now worse off than they would have been, because the only alternative would be a situation in which

---

4. See Kavka (1982) and Parfit (1982) for early but thorough discussions.

these individuals never come into existence. Thus we cannot say our choice to increase consumption wrongs, harms, or fails to respect the interests of any specific person, making it difficult to identify anything at all that is wrong with this choice. But intuitively it is *very* wrong for the current generation to make a decision that forces people 200 years in the future to live in deplorable conditions.

While various potential solutions to the Non-Identity Problem have been offered[5] (with the "exploitation" solution discussed in Human Societies in Space), for the moment I want to take the problem as a given. What does it imply for the duty to extend human life? Suppose we opt not to pursue space settlement, damning the human race to extinction sometime over the next few centuries or millennia. The problem is that, once humans no longer exist, there are no longer any persons who are harmed, wronged, or whose interests we have failed to respect. No one is worse off because of the expiration of our species, because in a different future with space settlement, entirely different persons would have come into existence. Thus we ought to reject in the first place that we have a duty to extend human life, via space settlement or otherwise.

Despite the apparent plausibility of this objection, the Non-Identity Problem is not genuinely applicable to the duty to extend human life. The duty to extend human life is not a duty to ensure the well-being or interests of any particular person in the future. Instead, this duty either is one which respects the interests of humanity as a whole, or is one which aims to promote the intrinsic value of human life (however that value is best conceptualized). Both offer ways out of Non-Identity issues. For instance, if human life is intrinsically valuable in something like Moore's sense,[6] then we can found an obligation to extend human life on an obligation to realize, promote, and maintain that which adds value to the world.[7] Alternatively, if we grant that humanity as a whole can have interests which ought to be respected, then we ought to respect humanity's interest in continued existence.[8] By saying that "we" are obligated to extend human life I mean to identify human persons *collectively* as bearers of this obligation. It is not individual persons, then, who have an

---

5. See Meyer (2016) and Roberts (2015) for overviews.
6. See Chapter 2, note 1, as well as my discussion in Chapter 4 of Charles Cockell's microbial ethics.
7. Such an obligation would have to be framed in a way that does not invite Derek Parfit's (1984) "Repugnant Conclusion," and this is no trivial task. For this reason, a Kantian approach to human intrinsic value may seem more attractive. Perhaps an obligation to ensure human survival could be derived from the "kingdom of ends" formulation of Kant's (1993) categorical imperative.
8. It is granted that we ought also to weigh the interests of the human species against the interests of its members. If the only way to extend human life is to force space settlers to suffer immensely and permanently, it might be better that humanity simply expires.

obligation to help extend human life (although we each have an obligation to *refrain* from actions that would do great harm to others or to the species as a whole). Rather, our obligation is to collaborate or work to create institutions and global arrangements that facilitate extending human life.[9]

Would it be wrong not to respect humanity's interest in continued existence? In failing to pursue space settlement, perhaps our omission would be a case of *permissibly letting die* as opposed to a case of *wrongful killing*. The trouble with this reply is that an essential feature of cases in which it may be permissible to let a person die is that the person in question will suffer unbearably if kept alive. It remains a clear case of wrongdoing to let a person die who could be saved easily, who expressly wishes to be saved, and who will afterward lead a worthwhile life. The case of extending human life, especially via space settlement, is much closer to this latter kind of case. Supposing then that space settlement can be carried out without detracting from other obligations, and that it will provide settlers with worthwhile lives, failing to extend human life via space settlement would constitute an impermissible case of letting die. Therefore, if we accept that humanity has an interest in continued existence, we should not deny the coherence of an obligation to extend human life via space settlement.

The Value of Humanity

If instead we opt to ground the duty to extend human life on the intrinsic value of human life, then we encounter another kind of criticism. If it turns out that human life is not intrinsically valuable, then no intrinsic good would be respected, promoted, or maintained directly by extending human life. Here it should be appreciated that the intrinsic value of human persons (or of features which humans possess such as sentience or rationality) tends to be assumed rather than argued for.[10] That is, it is a presupposition of modern ethical theorizing that human persons are intrinsically valuable, and it is as least logically possible that this presupposition is false. If pressed to defend this presupposition I would do so in a way that coheres with the approach to intrinsic value I outlined in Chapter 2, according to which intrinsic values can be inferred when something being valued for its own sake forms part of the best explanation of a practice that coheres with our overall scientific worldview. That humans are intrinsically valuable certainly passes this test: Professional ethicists seem never to doubt the intrinsic value of human life; and based on

---

9. See Tomalty (2014) for a defense of the related claim that humans have a collective obligation to provide to all persons the basic necessities of life.

10. But *cf.* Cicovacki (2017).

our sincere convictions and behavior we nearly universally value human life for its own sake, believe that humans should be treated as ends in themselves, and do not wish to see humanity expire.[11]

I confess that I am at a loss to justify the intrinsic value of human life to anyone who is not persuaded by evidence of the kind just mentioned. A dedicated skeptic might reject such evidence as a mere description of opinion or cultural bias (Traphagan 2019) and demand a more substantial, objective proof of the intrinsic value of human life. But answering the skeptic here would be a fool's errand, because the correct reply is to accuse the skeptic of asking for too much, or for more than is possible to provide. Ethical theorizing is inextricably intuition-laden, and intuitions about the value of human life—whether concerning the value of human life as such, or concerning the value of humans as sentient beings, or as rational beings, etc.—should not be dismissed as mere opinion when these intuitions survive the rigors of reflective equilibrium. That the intuition that human life is intrinsically valuable is propaedeutic to ethical theorizing suggests that it is theoretically and practically indispensable, and this is as secure of a position as one can hope for in ethics—or in any other domain, for that matter. So skepticism here, while perhaps impossible to defeat, is nevertheless insufficiently motivated.

Perhaps I have missed the point of being skeptical about the value of human life. Perhaps skepticism about the intrinsic value of human life is merely an accomplice to the laudable goal of overcoming human arrogance by helping us to realize that humanity is not *especially* valuable, and that we should be more appreciative of the intrinsic value both of other forms of life and of entities of other kinds. Here I would agree; there is good reason for us to become more humble in our attitudes toward ourselves and toward others (including nonhuman others). That, however, is insufficient to derogate a duty to extend human life. That human life is not the only source of intrinsic value does not imply that it would be wrong to extend human life, unless in the process more was lost than was gained. And while I will argue that this is the case over the short term, there is much less room for skepticism over the long term.

The upshot is that there are no plausible grounds for rejecting the duty to extend human life. While the duty may be overridden, it exists nevertheless. So, *ceteris paribus* and to the extent that space settlement is required for satisfying this duty, we ought to accept that we have an obligation to pursue space

---

11. On this point, then, it does not seem to matter whether we view human intrinsic value in Kantian or Moorean terms, since intuitions suggesting that humans are intrinsically valuable in both senses (assuming they really are two distinct notions) pervade normative ethical theorizing.

settlement. However, as I hope the next two objections will show, these two antecedent conditions are not met.

The Disposable Planet Mentality

According to a third objection, which is a kind of moral hazard argument, the *basic argument* has unacceptable consequences: Regardless of what happens in space, humanity retains obligations to, e.g., ensure the well-being of terrestrial humans, maintain terrestrial ecosystems, etc. If escape from Earth becomes possible, many humans may be much less motivated to help satisfy these obligations—they may adopt what William Hartmann (1984) calls the "disposable planet mentality." Lynda Williams articulates this concern succinctly:

> The hype surrounding space exploration leaves a dangerous vacuum in the collective consciousness of solving the problems on Earth . . . . Many young scientists are perhaps fueling the prophecy of our planetary destruction by dreaming of lunar and/or Martian bases to save humanity, rather than working on the serious environmental challenges that we face on Earth. (2010, 8)

Whether space settlement would actually encourage individual humans to develop the disposable planet mentality cannot be adjudicated without empirical research. It is also unclear how such an attitude, assuming it arises at all, would be distributed among humans. Possibly, the disposable planet mentality would only arise in extraterrestrial populations. Granted, it might not, and it would be most concerning if it were to take hold terrestrially—especially among terrestrial leaders. However, a number of practical considerations militate against extreme pessimism on this issue.

First, even if humanity pursues an aggressive space settlement program, only a very tiny fraction of the human population could ever hope to immigrate into space (at least in the absence of distant, futuristic technology). This means that for the foreseeable future, Earth will remain the only home for billions of humans. The vast majority of these terrestrial humans will have an interest in ensuring the continued ability of Earth to support worthwhile human lives. So even if extraterrestrial humans, preoccupied with mere survival in space, stop caring about the Earth, those for whom immigration into space is not possible will still be motivated to solve terrestrial problems. Second, those who immigrate into space will likely become *more* rather than less concerned about resource conservation. In space, all of the basic necessities of life must be created through artificial means. Wasteful or frivolous resource consumption in space, where there are no natural recycling systems, is significantly more lethal than wasteful or frivolous terrestrial resource consumption. Space societies may even come to adopt idiosyncratic and especially austere conceptions of what is wasteful (compared to their terrestrial counterparts). *A fortiori*, those who immigrate into space are unlikely to adopt the disposable planet mentality.[12]

What must be kept in mind, then, is that Earth is likely to remain an attractive place to live, and it is also likely to remain the most popular home for humans. These items, in conjunction with the harsh, instantaneously lethal environments of space, militate against the likelihood of a widespread adoption of the disposable planet mentality. For the foreseeable future, there will remain strong incentives for humans to solve terrestrial problems, even if some decide to immigrate into space. It is true that it seems unlikely that space settlement will directly help solve terrestrial problems, but that is not its primary function, which is instead to extend human life in the face of global terrestrial threats. (And it is again fair, for the near future, to question the wisdom of expending extensive terrestrial resources on space settlement if there is competition for these resources. We should not insist dogmatically that space settlement is always a priority; but nor should we insist, equally dogmatically, that it should never be prioritized.) The *basic argument*, then, survives the challenge posed by concerns about promulgation of the disposable planet mentality.

The Effectiveness of Space Settlement

One final objection to the *basic argument* is that it is false that human life *can* be extended via space settlement. For instance, many of the threats to terrestrial life remain threats to life throughout the Solar System: asteroid collisions; war; disease; etc. Space settlements will not *eliminate* existential risk for any group of humans. (This is especially true over the very, very long term, since no place in the Universe is habitable permanently. Still, we have trillions of years before the "Heat Death" renders the entire Universe uninhabitable.)

Here again we might accuse the skeptic of asking for too much. It is granted that space settlements, including interstellar settlements, will not provide permanent and absolute guarantees for extending human life. But with each new self-sustaining settlement the risk of *total* human extinction diminishes. So the obligation to extend human life via space settlement is, fundamentally, an obligation to *minimize the risk* of total human extinction. And, as most often promulgated, it is an obligation to minimize the risk of total human extinction due to global terrestrial catastrophe. Permanent, self-sustaining space

---

12. See Munévar (2014; 2016) for an alternative response to environment-oriented objections to space exploration and settlement.

settlements are perfectly capable of reducing both of these risks. But are they worth the effort? Would the risk reduction provided by space settlements be sufficient to justify the vast monetary, physical, and human expenses needed in order to establish these settlements?

On timescales measured in decades to one or two centuries, self-sustaining space settlements, should any be established, are likely to have small populations and to grow less quickly compared to historical European colonies in North America. Not only could few humans afford the trip, the difficulties inherent in producing basic, life-giving resources such as water and oxygen in the inhospitable environments of space will significantly limit the rate at which space settlements can grow in population and extent. This means they will be less robust, more tenuous, and much more likely to collapse when compared with mature terrestrial societies. So, over these timescales they would not provide a significant reduction in the risk of total human extinction. It is true that a mature, self-sustaining space settlement would provide a significant reduction in risk, and that such a settlement would probably develop out of very humble beginnings—that is, it is true that we would have to start *somewhere* and *at some time*. But that does not mean we have a compelling reason to start anytime soon (or to devote much in the way of current societal resources in preparation for later space settlement), especially given the existence of other more achievable means for mitigating many existential threats. A more effective method for protecting humanity against the threat of asteroid collisions, for instance, would be to increase our asteroid cataloging efforts and devise mature technologies for asteroid diversion. Doing these things would have the added benefit of protecting all of the life on Earth, whereas settlement, if pursued as the exclusive means for extending human life, would not help any terrestrial humans living in the path of an asteroid. Neither option has to be pursued exclusively, but resources are limited. Protecting Earth, as opposed to settling space, is likely to be a cheaper and a more effective insurance policy against humanity's expiration due to an asteroid impact.

A massive expenditure in support of *urgent* space settlement would therefore be difficult to justify.[13] Nevertheless, as threats to terrestrial life increase, and as humanity expands its technological capabilities, it becomes easier to justify space settlement attempts, which become both less expensive and more effective at reducing the risk of total human extinction. So the ineffectiveness objection, while it misfires against the *basic argument*, nevertheless hits the mark when it comes to the *urgent argument*.

The *basic argument*, therefore, survives the brunt of criticism. We must accept that we have a duty to extend human life. Furthermore, we must accept

13. See also Milligan (2011).

that space settlement is, in principle and over the long-term, required for satisfying the duty to extend human life. To paraphrase Kant, since those who will the end will the means, it follows that we have an obligation to pursue space settlement. This duty is not urgent, as the ineffectiveness objection unveils. As I argue in the next section, this judgment is reinforced by the potential for space settlement to conflict with space science.

AGAINST THE URGENT ARGUMENT

That practical considerations speak against the *urgent argument* serves to undermine the effective strength of any present-day obligation to pursue space settlement. This does not defeat entirely the argument, since in the absence of any countervailing considerations we would still to some small degree have an obligation to work towards space settlement. What the relative weakness of the *urgent argument* allows is that an *urgent* duty to extend human life via settling space can be rather easily overridden should it come into genuine conflict with other duties, even if the latter aspire to something of less importance than extending human life. As should at this point come as little surprise to the reader, I propose the duty to increase our scientific knowledge and understanding of the universe as one such conflicting, overriding duty. Although the duty to acquire new scientific knowledge and understanding may be weaker when compared with the duty to extend human life, nevertheless the duty is stronger in practice. This is because the former can be satisfied much more effectively and at a significantly lower cost.

But is the conflict between space science and space settlement genuine? In most cases the answer is "Yes." As I argued in Chapter 4, space science has yet to disclaim interest in any particular space environment, motivating a precautionary principle according to which a space environment should be presumed to be of scientific interest until proven otherwise. And in Chapter 5 I used this precautionary principle to reject an urgent duty to exploit space resources. The same basic strategy suffices here: Space settlement would in most cases result in the disruption or destruction of a pristine space environment, significantly reducing (or vitiating entirely) the scientific value of this environment. This is most evident in the case of destinations of interest in the search for extraterrestrial life, most notably Mars. Thus, I share Ian Stoner's worry that "[a]ny human presence on Mars is likely to constitute a significantly invasive or destructive investigation of the Martian environment" (2017, 344). John Rummel et al. elaborate:

> If humans are intended to forward astrobiological objectives, it would be impossible to achieve them if abundant Earth life were to be introduced into every

environment visited on Mars—and disastrous if the crew were unprepared to deal with martian life if they find it. As such, planetary protection provisions for future human missions to Mars are essential to mission success, and must be integrated into such missions during the earliest stages of their design and development. Spacecraft and planetary habitat designs must accommodate those provisions, and the microbial ecosystems within the spacecraft and habitats on Mars should be well known and monitored carefully throughout the mission. (2010, 165)

There are already considerable concerns about the biological contamination of Mars from robotic exploration; these concerns would be magnified considerably by crewed exploration, and magnified considerably more by permanent Mars settlement. A Mars settlement might preclude entirely our ability to establish decisive conclusions about the existence or not of endemic Martian life, as Rummel explains in a more recent paper:

> [O]n Mars even the most basic attempts at local or later planetary-scale engineering can result in changes to the environment that could greatly affect the ability of Earth organisms to thrive there, as well as the prospects for interaction with any martian organisms that might be present. (2019, 626)

For instance, Mars's atmospheric methane—which is a potential biosignature—rarely exceeds 1 part per *billion* (Webster et al. 2018), meaning even a modest human presence could release enough methane to interfere with a study of the natural methane flux. Similarly, exploitation of Martian water sources, including the polar layered deposits as well as the ground ice that exists under roughly 1/3 of the Martian surface (Dundas et al. 2018) could hinder the study of Mars's climate history (destroying clues to its past habitability)—and of course, frozen in the ice there might exist evidence of Martian life.

But the risks to science posed by space settlement are not limited to astrobiology and the search for extraterrestrial life. As mentioned in Chapter 5, disruption or destruction of any space environment compromises its scientific integrity. Activities associated with space settlement—spacecraft landings and takeoffs, habitat construction, in situ resource utilization, etc.—would do irreparable harm to non-biological forms of exploration. On Mars (which is the third most-explored body in our Solar System, after Earth and the Moon), nearly all climatological, hydrological, and geological processes are unconstrained (Banfield 2018). This means that we know very little about what kinds of processes take place, how they vary over different timescales, etc. We lack sufficient understanding to predict the nature and scope of the consequences of a permanent human presence on Mars. So, not only can we not yet measure settlement's cost to science, we cannot even predict whether certain forms of resource exploitation (such as drilling into the surface to melt ground ice for a water supply) could be conducted sustainably. For various reasons, then, we ought to refrain from placing human settlers in positions where they must, in order to survive, destroy or exploit the resources of an unexplored environment.[14]

My position raises a question about how it is to be decided when to first allow space settlement. After all, it is not possible to remove all doubt surrounding a space environment's potential interest to science. And it would be foolish to decide right now what that threshold is. But as a reminder, virtually every space environment is substantially unexplored. Space science *in toto* has yet to disclaim interest in any particular space environment, planetary body, moon, asteroid, comet, etc. So at the present time, it is appropriate to adopt as a precautionary default that all space environments are worth preserving for scientific study. Moreover, we should not be so foolish as to presume that the concerns of present-day science will be the same as the concerns of science 20, 50, 100, or 500 years from now. We would do better by humanity if we provide future scientists with more, rather than fewer, possible objects of study. This means we should not, in our ignorance of the future needs and interests of science, engage in wanton destruction or disruption of space environments, whether for the purpose of settlement or exploitation (or both).

Second, we should recognize the continuing importance of science as a stakeholder in decisions about space settlement. There is an asymmetry of costs: Preserving an area for scientific study is compatible with future settlement in this area; whereas settling in some environment could preclude the effective study of this environment. Thus, decisions about space settlement destinations should include substantial input from representatives of the scientific community, so that settlement activities can be halted if they appear likely to disrupt a site of sufficient apparent value to science. Moreover, those interested in settling space will very much *benefit* from the results of scientific study. It would be irresponsible to settle in a space environment that had not been the subject of prior, intensive scientific investigation. Otherwise potential emigres could not be confident that this environment was capable of supporting a self-sustaining human community. Such acts of preservation need not interfere with our long-term duty to extend human life via space settlement, especially if early on we prioritize the scientific exploration of likely settlement destinations. As in the science-versus-exploitation dispute, the conflict between the duty to scientifically explore space and the duty to settle space is temporary. And although we cannot predict precisely when that conflict will dissolve, there is every reason to expect that its dissolution will

14. For further discussion of how settlement threatens Mars science, see Schwartz (2019c).

come quickly enough to satisfy our obligation to extend human life via space settlement.

HUMAN SOCIETIES IN SPACE

Given the cogency of the *basic argument* and that the conflict between space science and space settlement will eventually dissolve (perhaps more quickly in the case of lunar science-versus-settlement than in the case of Mars science-versus-settlement), I accept that we ought to someday undertake the settlement of space. When that time comes, however, we must grapple with ethical questions of an entirely different sort. Space settlement, at least as it is usually described, requires a group of people to consent to a one-way trip to some destination in space—a voyage which might take months and which they may not even survive. If they survive the outbound trip they will then struggle to construct and provision a self-sustaining habitat that protects them from the instantaneously lethal environment of space, which likely will afford them scant privacy or freedom of movement. Eventually settlers will have to rear children, who may have no control over the restricted conditions of life found within their settlement. Is this a kind of enterprise that we ought to encourage? Or even permit in the first place? Two questions come to the fore: What would be required for potential emigres to give their voluntary, informed consent to instigate space settlement? And would it be permissible to subject future generations of settlers to the conditions of life within a space settlement?

Consenting to Space Immigration

By default we ought to respect the autonomous decisions of informed, rational agents. If a person were to genuinely understand the personal risks and costs associated with space immigration, and were they to process this information rationally, we ought not to object should they decide to take part in early space settlement initiatives.[15] There are many unknowns regarding the risks of near-term crewed spaceflight initiatives, especially given recent interest in space tourism and in what medical or other requirements might exist for "spaceflight participants" as opposed to professional astronauts (Langston 2016). There are even more unknowns when it comes to the risks associated with *permanent* space habitation. We do not presently know whether human

---

15. Traditional accounts of voluntary, informed consent do not necessarily require the rational processing of information, but Savulescu and Momeyer (1997) argue convincingly that full autonomy requires the use of theoretical rationality.

beings can survive long-term, reproduce, and flourish in either reduced gravity or microgravity environments. Thus it would be highly premature to say that we have an adequate scientific understanding of the personal risks associated with permanent space settlement.[16] But to be fair, what is under discussion here is not the kind of urgent, learn-as-you-go approach to settlement championed by organizations like Mars One, but instead a measured approached in which settlement is not attempted for several decades or centuries. As a charitable simplifying assumption, then, let us imagine that there are no genuine physiological hurdles to human space settlement.[17] This allows us to sidestep the serious objections that David Koepsell (2017) has raised about consenting to near-future Martian settlement.

The main concern that remains, then, can be seen best by analogy with consenting to experimental medical procedures. Intuitively, consenting to an experimental procedure—where little is known about possible risks and their likelihoods—requires more careful deliberation than is required for consenting to well-established, effective treatments. Decisions to consent to experimental procedures are often made under duress, where participants judge that the risks associated with unknown effects are less severe than the risks associated with alternative treatment options. Those who consent to experimental medical procedures have a reasonable hope that after the procedure they will be better off than they would have been if they had decided otherwise. Given that space travel is likely to remain considerably expensive, those for whom space immigration is possible will be fairly affluent individuals. It is therefore likely that their lives would be made considerably worse off by immigrating into space, which would not allow them to exercise many of the freedoms associated with terrestrial living (not having to toil constantly to produce basic resources; being able to walk outside without a cumbersome pressure suit; freely associating with any of a very large group of individuals; etc.). Their decision to immigrate into space would be analogous to a perfectly healthy person requesting an experimental procedure that would provide them with no possible medical benefit and would more than likely leave them worse off. Respect for personal autonomy may ultimately require us to cede to the wishes of such persons. But whatever we ultimately say, our standard clearly ought to be *higher* than in the case of the chronically ill who consent to experimental procedures. A person's mere expression of a desire to immigrate

16. For an overview of current understanding see Nicogossian et al. (2016).
17. Since all evidence so far suggests that humans do not adapt terribly well to the space environment, the overcoming of any physiological hurdles associated with permanent space settlement would require extensive medical experimentation and study. This raises thorny questions about the ethics of using human subjects in very long-term settlement experiments—questions I will have to pass over here.

into space is insufficient. Great care must be exercised in order to be certain that this person *genuinely understands* the risks of space settlement; that they can *deliberate rationally* about these risks and draw reasonable inferences from them; and that these risks are compatible with their overall values and life ambitions. The process should not be an easy or a simple one.

Future Generations of Settlers

Imagine, then, that a group of potential emigres have unproblematically consented to immigrate into space. This would include consenting to living with only a small number of other settlers, which would severely limit their opportunities for sexual and romantic companionship. This would also include consenting to living a life that is restricted in many other ways: Vital, life-supporting technologies must be provisioned and maintained; essential skills must be kept up—thus these emigres will have little say in their selection of vocation. Each must do what is required for the survival of the settlement. Settlers might even have to agree to live under a very illiberal social organization, as Charles Cockell has warned:

> The sheer extremity of space compared to Earth lies in the unrelenting requirement to artificially produce water, air and food regardless of one's location. These requirements generate an intricate interconnected labyrinth of life-support systems upon which everyone depends for their survival. In essence, the instantaneously lethal conditions in outer space require a thorough and unfailing requirement for the servicing and maintenance of the complexity of machinery that enables life. (2016b, 24)

He expects that settlers would demand a culture of conformity, one with severe disincentives for dissent, lest new and dangerous ideas or activities pose a threat to life-support and maintenance systems. This leads to a significant risk of the emergence of autocratic or totalitarian forms of government, according to Kelly Smith:

> If you thought it a simple matter for someone to, without warning, destroy everyone and everything you cared about, what would you do? Recent history seems to indicate that people are in fact quite willing to give up hard won freedoms when they feel sufficiently threatened, even in a mature democratic culture. So it seems quite possible that colonists in such conditions would *demand* repressive measures such as . . . pursuing invasive surveillance of the population. In other words, the very same aspects of a colony that make it worryingly easy to exert totalitarian control could also make such control *adaptive*. (2016b, 180)

And we might well accept that an informed rational agent could freely consent to living such a life. But to be a truly *permanent* and *self-sustaining* settlement, the settlers eventually will have to rear children. Their children will neither have consented nor could consent to living in a space settlement, and might never be given the option to leave, as the settlement may not have the resources to construct and provision return trips to Earth. What's more, children reared in reduced gravity environments might not be physiologically capable of adapting to Earth-standard gravity. So even if a settlement had the means to transport unhappy denizens back to Earth, it may not be possible for these individuals to survive there. It is far from obvious that settlers should be permitted to subject future generations to these restricted conditions of life. Thus we can no longer avoid the question of duties to future generations.

Recall that the Non-Identity challenge to duties to future generations arises because decisions made at one point in time affect who comes into existence later on. In the case of space settlement, the problem is that we could not say that children born into a space settlement would be harmed by the heavily restricted, confined environment into which they are born. This is because had their parents not immigrated to space, or had space settlement not taken place until technological and other improvements provided settlers with considerable personal liberties, these specific children would never have been born. Hence, they would not be "worse off" on account of being born and having to live in an environment providing few personal liberties. A possible way to block the Non-Identity problem, and thereby save the intuition that it would be wrong to do this to a child, is to locate a non-harm-based wrongmaking feature of such an action. One attractive option here is to argue that such an action would be wrong because it would be unjustifiably *exploitative*. Traditionally, however, exploitation involves harm, as typical examples include: severely underpaying employees whose only alternative is unemployment; charging an exorbitant fee for immediately necessary resources or assistance, and so forth. In such cases exploiters *unfairly benefit*, while the exploited are *unfairly harmed* or disadvantaged, either because they must act out of urgency or with severely restricted options. However, it is possible to show that exploitation occurs even when the exploitative transaction *benefits* the exploited party.

Consider the following case discussed by Hallie Liberto (2014) and due originally to Valdman (2009):

> *Antidote Case*: Hiker A carries with him an antidote for a deadly poison contained in the venom of a snake living in the woods. Hiker A encounters Hiker B who has just been bitten by such a snake and who has no antidote with him. The market value of the antidote is $10. However, A knows that B is worth a million dollars

and, so, charges him a million dollars for the antidote. B unhappily accepts the terms of the transaction, uses the antidote, and lives. (Liberto 2014, 75)

In this case, both the exploited and the exploiter are made better off: Hiker A is better off, for now he has acquired nearly one million dollars. Hiker B is also better off, because now he is alive rather than dead. Still, Hiker A wrongfully exploits Hiker B. This case, unfortunately, is not perfectly analogous to the relationship between space immigrants and their potential offspring. Space immigrants, living in the same poor conditions as their offspring, do not obviously benefit by procreating. But as two further cases show, neither the existence of a benefit nor the anticipation of a benefit is necessary for exploitation:

Antidote Case B: The same as the original Antidote Case, only instead, "Hiker A demands the impressive looking ring off of Hiker B's finger in exchange for the life-saving antidote. Though the ring has sentimental value to Hiker B, it is actually worth nothing on the market. Hiker B tries to inform Hiker A of this fact, but Hiker A does not believe him." (82)

Antidote Case C: The same as the original Antidote Case, only instead Hiker A demands that "Hiker B write out a check for $100,000 to her son in return for the antidote." (82)

Antidote Case B supports the intuition that exploitation can occur even when the exploiter does not benefit; Antidote Case C supports the intuition that exploitation does not even require the *expectation* of a benefit. One might complain that in Antidote Case C Hiker A still receives a benefit—for instance, the satisfaction they experience in knowing their son is about to receive a significant windfall—and that this benefit is essential to the wrongfulness of the exploitation. Such an objection can be overcome easily. Suppose instead that Hiker A planned to cash the check and then to burn the cash—without deriving any benefit from those activities. Or, in a case more clearly relevant to extending human life via space settlement, suppose that Hiker A demands a check for the same amount written to the Sierra Club, or to Greenpeace, or to some fund established to help pay for conservation efforts benefiting future generations. Suppose further that Hiker A makes this demand feeling little to no concern for environmental issues or the well-being of future generations. These are still cases in which Hiker B is exploited.

As Matthew Rendall argues, what is wrong in such cases (even though the exploited individual is better off as a result of the transaction) is that "[t]he exploited party receives less than she deserves according to the relevant 'fairness baseline'" (2011, 237). His mention of a "fairness baseline" indicates that, in order to avoid exploitation, it is not enough that we merely ensure that transactions do not result in individuals with lives that are not worth living. As he notes, what we can and should object to in certain cases is that the outcome for exploited individuals "should have been better, not that it is absolutely bad" (236). We find fault in actions which create situations in which *avoidable* harms or disadvantages occur, and in which these harms or disadvantages are not compensated for in a reasonable way. In the case of space settlement, the relevant harms or disadvantages could be avoided either by foregoing space settlement, or by waiting to settle space until the maturation of technologies that permit freer lives within settlements.

In this regard I do not view vocational or educational restrictions as the most troublesome. Rather, what is most potentially exploitative and therefore most concerning are the severe restrictions on reproductive autonomy that are likely to occur within space settlements. Given the great expenses associated with space travel, early space settlements are likely to contain the smallest feasible populations. Most estimates for minimum viable populations (respective of genetic diversity as well as diversity of skills) range between 40 and 100 individuals.[18] In such small settlements (assuming they remain relatively isolated) it would be imperative for long-term survival that each settler participates in highly regimented, planned reproductive activity, whether or not they have any desire to do so. But certain individuals may wish not to reproduce, or at least, may not desire to do so with any particular member of their settlement. There is a word that describes the act of forcing an individual to participate in reproductive activities against their will, *viz.*, "rape." Rape is one of the few absolute wrongs; at no time should it ever be condoned, and at no time should we encourage or help to create a society that cannot persist without rape or that would normalize or be likely to eventuate in rape on a wide scale.[19] It is incredibly alarming, then, to find space settlement advocates reacting nonchalantly to the potential for space settlement to result in generations of rape, often rationalizing the problem as an unfortunate, though tolerable cost of space settlement. While discussing reproductive autonomy in space settlement, Konrad Szocik et al. suggest that "psychological training of astronauts should include preparation to withdraw moral intuitions

18. See Moore (2003); Cooper (2016); Marin and Beluffi (2018); Mankins, Mankins, and Walter (2018). A rare exception is Hein et al. (2012).

19. *Cf.* Tony Milligan on bodily integrity and abortion in space: "Bodily integrity is in any familiar liberal context, non-negotiable. It is basic to our conception of respect for the individual, indeed it is integral to our terrestrial understanding of what it is *to be* an individual" (2015c, 16). "The very fact that abortion rights might then have to be compromised, and perhaps indefinitely so, if a settlement is to survive . . . could shape our attitude towards how worthwhile the creation of a space settlement *in such places* might be and whether or not we terrestrial agents should ever support it and thereby condemn future humans to a partly-illiberal predicament. The creation of any settlement with a built-in requirement for norms which are radically different form our own would be a serious matter and might be difficult to justify" (20).

that—even if deeply rooted—may need to be revised to build a new world" (2018, 61). One sincerely hopes they do not view the belief in the absolute wrongness of rape as a mere intuition in need of revision if it proves antithetical to pursuing space settlement! Similary, *sensu* (Smith 2016b), perhaps a tolerance for heavily coerced reproduction could become adaptive for space cultures. To this the correct reply is that there are some conditions to which no person or society should be forced to adapt, and among these include conditions which would make heavily coerced reproduction adaptive.

A more measured approach, then, would admit of a lower bound on the size of a space settlement, so that it remains large enough to sustain a genetically diverse population solely through voluntary acts of reproduction. But would a space settlement be large enough if it *barely* avoided rape as a requirement of its persistence? According to one of the most influential—and as I shall argue, misguided—papers on the related topic of interstellar settlement, the answer is "Yes."

Interstellar settlement presents a unique challenge: The closest potentially habitable exoplanet, Proxima Centauri b, is over four light years (about 40 *trillion* km!) distant. Even traveling at 1% of the speed of light—a feat well beyond envisioned capabilities for crewed missions—it would take over 420 years to reach Proxima Centauri b. Most other potentially habitable worlds are more distant by at least an order of magnitude. For this reason many have proposed interstellar settlement by means of large, multigenerational "worldships." Aboard these vessels, numerous generations would have to be brought into existence, with these individuals—the shipborn—having no choice except to spend their entire lives confined to their worldship. According to Edward Regis (1985), there is nothing obviously problematic about subjecting the shipborn to the conditions of life aboard a worldship. For were we to invoke the intuition that it would be wrong to subject the shipborn to these conditions, then this "would make the preponderance of historical and perhaps even contemporary procreation immoral, a conclusion probably not acceptable to many of those who procreated in those conditions, nor to their offspring" (255). Thus, the permissibility of procreation, past and present, in significantly non-ideal settings suggests that procreation aboard a worldship would similarly be permissible. Furthermore, he insists, it would not be wrong to do this because it does not constitute an example of a rights violation:

> Will there be enough people aboard for variety in human companionship, for adequate choice in selecting a romantic partner, or for a stimulating and well-developed culture? But however we may answer, we must understand that here we are no longer speaking of that to which anyone could justly claim a right. For although these conditions may be necessary for an ideally fulfilling life, it does

not seem that people have rights to such lives nor therefore to the conditions that would make them possible. (255)

It follows that we should not object to worldship travel on account of the restrictions it would place on the shipborn.

There are three significant problems with Regis's defense of worldship travel—problems that would also arise were Regis's arguments used in support of more local forms of space settlement. First, moral obligations are not limited to refraining from committing rights violations. Even if it is true that no one has a right to, e.g., adequate choice in selecting a romantic partner, it does not follow that we do nothing wrong by depriving people of such choices. Second, Regis appears to be working with an unnecessarily limited conception of rights. It may be true that there is no *negative* right to adequate choice in selecting a romantic partner, but perhaps there is a *positive* right to such choice. Such a case might be made as an extension to defenses of subsistence and other positive rights.[20] In either case Regis goes wrong by focusing exclusively on obligations to avoid harm, while failing to recognize, e.g., duties to care. And once we recognize a duty to care, the space settlement question transforms from "What should we do to prevent the settlers/shipborn from suffering lives that are not worth living?" into "What should we do to ensure that the settlers/shipborn live the best possible lives?" A defensible answer to this question, especially as it pertains to romantic companionship, must face up to certain realities: Even if each initial settler or worldship passenger is heterosexual and willing to reproduce for the good of the mission, among their children there inevitably would be persons who are homosexual, bisexual, and asexual, as well as persons who are agender and transgender. Moreover, there would be people who want children and people who do not want children. It would be difficult to justify forcing any such individual to live their entire life in an environment that did not provide them with the opportunity to pursue (or not) the romantic relationships of their preference.

Third, Regis's framing of the worldship question is misguided from the outset. He approaches the issue from the point of view of prospective parents aboard a worldship who are considering whether it would be permissible to rear the next generation. As applied to near-term space settlement, we might envision the first generation of Mars settlers debating whether it would be justifiable to subject future children to the conditions of life on Mars. And here is where the analogy with past and present procreation under non-ideal conditions is genuinely persuasive: I would agree with Regis that, e.g., parents are usually blameless for any harms or disadvantages their children experience due to being born in significantly non-ideal conditions that are created

20. See, e.g., Shue (1996) and Ashford (2007).

through forces outside their parents' control (at least as long as they have a reasonable expectation that their childrens' lives will be worth living, even if only barely so). Prospective parents living in despotic regimes should not be prohibited from having children, and should not be blamed for the conditions their children must endure. A similar judgment is appropriate for prospective parents living in an already established space settlement. But the question here of whether to pursue space settlement is not whether procreation is permissible in *already established* conditions of hardship. Instead, the question of whether to pursue space settlement is a question about whether it is permissible to *create the conditions of hardship* that would impact many generations. Thus, the relevant analogy is not with the deliberations of prospective parents living in despotic regimes. Instead, the correct analogy is with the *despots* who are deliberating about how to treat their citizens. And certainly despots who oppress and exploit their citizens act very wrongly. Similarly, we would act very wrongly if we *founded* a new society in which we expected its citizens to enjoy few liberties, to experience severe hardships, and lead limited lives. This would be unjustifiably exploitative.

While some lines ought never to be crossed, e.g., we ought never permit a space settlement that requires rape as a condition of its persistence, nevertheless as Liberto rightly reminds us, most wrong acts, including acts of wrongful exploitation, are *prima facie* wrongs which might in certain circumstances become justifiable, all things considered (2014, 83). If the exploitation of space settlers can be kept to a minimum, space settlement becomes easier to justify. A space settlement that afforded settlers with considerable opportunities for romantic companionship would be easier to justify ethically than one which did not provide settlers with these opportunities (and likewise for educational and vocational opportunities). The overall good realized by having a space settlement matters as well. And since a permanent, self-sustaining space settlement would increase the chances of long-term human survival, this benefit must be factored into any decision about whether to initiate space settlement. But due to the failure of the *urgent argument*, the survival benefits of hastily organized space settlement are not especially great, and consequently they are unlikely to outweigh the significant human costs which space settlements are expected to exact. We should therefore exercise caution in settling space, and make sure to only attempt to create new societies in space when we can be confident that settlers will have substantive opportunities for realizing the ideal, liberal life.

Epilogue

Fate Amenable to Change

To close this book, I would like to describe how science will remain important even in futures (near or otherwise) in which humans have established space settlements. I would also like to address what can or should be said about the justification for space science if we relax the conservative assumptions used throughout the main text—a restriction to consideration of the "near future" (i.e., a time horizon including the present through the next two centuries), and the absence of major technological advances that could increase spaceflight activities by more than one or two orders of magnitude.

THE ENDURING SIGNIFICANCE OF SCIENCE

To repeat a point from Chapter 6, scientific examination of space environments is needed in advance of settlement to establish the viability of these environments for human habitation. Nevertheless, even in established space societies there will be a need to support research to assist in the maintenance of the settlement's life-support systems; to more efficiently acquire essential resources; and to improve the overall conditions of life for space-dwellers. Indeed, this need is likely to be greater for space societies than it is for terrestrial societies. Therefore, the justification I provided in Chapter 3 for the state support of scientific exploration and research applies, and to an even greater degree, to societies in space. The value of scientific research will not diminish as humans eventually pursue space settlement; rather, it will increase dramatically.

There are many hurdles to the creation of a space society. Space settlement is neither an urgent duty nor one that we could satisfy effectively even if it were required urgently. It is something we should pursue eventually, but only after we have conducted extensive scientific research of the space environment. And only after we have acquired the requisite technical and physiological knowledge for ensuring a very high probability of success, and a very high probability that settlers and their descendants will lead meaningful lives and will not experience excessive coercion in their decisions about what they would like to learn, how they would like to contribute to society, and with whom they will share their affections. These hurdles work together to reinforce the essential need to engage in scientific exploration and research during all stages of space settlement.

As explained above and in Chapter 6, intensive scientific investigation of the space environment is needed prior to settlement. This is not simply because of the intrinsic value of any knowledge or understanding generated by this investigation, and not simply because settlement would diminish the scientific value of various locations in the Solar System. Scientific investigations are also needed for the more practical reason that without them we could not be confident that a particular space environment was suitable for permanent human habitation. Scientific research, then, is an important enabler of space settlement. But planetary science and astrobiology are not the only relevant disciplines. In so far as the goal of space settlement is the production and maintenance of a society of *human persons*, extensive physiological, psychological, sociological, anthropological, economic, political, and philosophical work is needed on the various problems raised by space settlement. With respect to reproductive autonomy and opportunities for romantic companionship, psychology and sociology are especially vital. We need to know, for instance, how large a population must be in order to accommodate the companionship needs of a population comprised of more than just cisgendered heterosexuals.

Research will remain important after settlements are up and running. Indeed, much of what we today describe as "space science" will to settlers simply be called "science." To help appreciate the significance of science to a space society it is worth briefly rehearsing the instrumental justification for science that I provided in Chapter 3. The basic point underlying this discussion is that scientific exploration and research are vital and productive means for realizing social goods. Science is incredibly instrumentally valuable for human societies. Moreover, research of many varieties is necessary for healthy democratic deliberation and governance. Scientific research helps resolve political questions, helps to maintain the material preconditions of the democratic process, and helps support democratic culture. The result is that democratic states are obligated not only to protect freedom of *ends* of scientific research, but also freedom of *means* of research. Thus, democratic states have an obligation to provide wide-ranging support for scientific research. This reliance on science will be even greater for democratic states in space. Space societies, even if they are not democratic, will experience extreme challenges in providing for the material needs of their citizens—shelter, water, food, air, etc.—needs which can be so readily met on Earth, but not in space. Consequently, a democratic state in space would experience a comparatively stronger obligation to support scientific research—especially research that promises to help this society maintain the material preconditions of democracy. Moreover, all space societies, regardless of their form of governance, will share a compelling interest in supporting research that is necessary for developing and improving technologies required for securing the basic provisions of life. The particulars will vary from destination to destination. The understanding and skills needed to support a lunar settlement would not be exactly the same as those needed to support a Martian settlement, or an asteroid-based settlement.[1] Despite these differences, the common need to engage in effective research remains.

It will also be very important to provide space settlers with opportunities and resources for satisfying intellectual curiosity, for creating and enjoying aesthetic experiences, etc. Consider living in a lunar settlement. In contrast to life on Earth, life would probably seem extremely dull and confining. Even if the average citizen had the ability to venture beyond the confines of the settlement, they would be confronted with the rather bleak and homogeneous lunar surface. (Which admittedly may not appear all that bleak or homogenous to those reared on the Moon.) Charles Cockell proposes that in order to escape from this "extraordinary banality in the day-to-day experiences of weather, climate, and general environmental changes," we must encourage "intellectual efflorescence" (2008b, 270). I would make an even stronger claim: Without such opportunities for intellectual efflorescence, life in a lunar colony would hardly be worth living. That scientific freedom can create such opportunities counts strongly in its favor, with a similar defense supporting research into the arts and humanities.

There is, then, an enduring and omnipresent need to support scientific research. Scientific research, including space science, provides us with intrinsically valuable knowledge and understanding. Moreover, it provides us with a reliable means for improving society by solving existing problems, as well as by anticipating and devising solutions to new problems. This holds whether human society remains on Earth or extends into the space environment.

1. This might also imply that Mark Brown and David Guston's (2009) prioritization of "political rights" and associated research is not universally appropriate, since "social rights" (those associated with maintaining the preconditions of democracy) may be of the highest importance in space societies.

FUTURE IMPERFECT

What should be said, then, about timescales extending beyond two centuries into the future? And what should be said if revolutionary technologies open up the space environment in ways that render considerations of cost and scarcity moot? With respect to longer timescales, it could well be that lunar and Martian settlements become possible with minimal costs to science—especially if effective research is conducted over the near future. But this does not automatically "open up" the rest of the Solar System to exploitation and settlement. In particular, if no extensive efforts are made to explore the Main Belt asteroids and the outer Solar System—Jupiter and its satellites; Saturn and its satellites; etc.—then exploitation of or settlement in these places would again occur only at a significant and possibly unacceptable cost to science. Thus, while something like a 200-year moratorium may be appropriate in the case of Martian exploitation and settlement, longer moratoria are likely necessary for other destinations in our Solar System.

With respect to the possible development of revolutionary technologies, if spaceflight becomes relatively inexpensive, then space science costs should also decrease dramatically. This means that the opportunity costs associated with prioritizing scientific exploration should decrease as well, because more scientific exploration missions can be conducted and at a much quicker pace. So, even in a near future that includes fast, inexpensive space travel, it does not follow that the private sector should be given free rein to exploit and use space resources however it would like. Regardless of how we vary possible details about humanity's future in space, the value of science in space exploration persists.

APPENDIX A
Degree Conferral and Federal Funding

Data sources for the tables in this appendix:

- National Center for Education Statistics, Digest of Education Statistics. Each accessed 20 March 2018.
  - Table 325.10. WEB-ONLY TABLE—Degrees in agriculture and natural resources conferred by postsecondary institutions, by level of degree and sex of student: 1970–1971 through 2015–2016. https://nces.ed.gov/programs/digest/d17/tables/dt17_325.10.asp?current=yes.
  - Table 325.15. WEB-ONLY TABLE—Degrees in architecture and related services conferred by postsecondary institutions, by level of degree and sex of student: Selected years, 1949–1950 through 2015–2016. https://nces.ed.gov/programs/digest/d17/tables/dt17_325.15.asp?current=yes.
  - Table 325.20. WEB-ONLY TABLE—Degrees in the biological and bio-medical sciences conferred by postsecondary institutions, by level of degree and sex of student: Selected years, 1951–1952 through 2015–2016. https://nces.ed.gov/programs/digest/d17/tables/dt17_325.20.asp?current=yes.
  - Table 325.25. WEB-ONLY TABLE—Degrees in business conferred by postsecondary institutions, by level of degree and sex of student: Selected years, 1955–1956 through 2015–2016. https://nces.ed.gov/programs/digest/d17/tables/dt17_325.25.asp?current=yes.
  - Table 325.30. WEB-ONLY TABLE—Degrees in communication, journalism, and related programs and in communications technologies conferred by postsecondary institutions, by level of degree and sex of student: 1970–1971 through 2015–2016. https://nces.ed.gov/programs/digest/d17/tables/dt17_325.30.asp?current=yes.
  - Table 325.35. Degrees in computer and information sciences conferred by postsecondary institutions, by level of degree and sex

of student: 1970–1971 through 2015–2016. https://nces.ed.gov/ programs/digest/d17/tables/dt17_325.35.asp?current=yes.

- Table 325.40. WEB-ONLY TABLE—Degrees in education conferred by postsecondary institutions, by level of degree and sex of student: Selected years, 1949–1950 through 2015–2016. https://nces.ed.gov/programs/ digest/d17/tables/dt17_325.40.asp?current=yes.
- Table 325.45. Degrees in engineering and engineering technologies conferred by postsecondary institutions, by level of degree and sex of student: Selected years, 1949–1950 through 2015–2016. https://nces. ed.gov/programs/digest/d17/tables/dt17_325.45.asp?current=yes.
- Table 325.50. WEB-ONLY TABLE—Degrees in English language and literature/letters conferred by postsecondary institutions, by level of degree and sex of student: Selected years, 1949–1950 through 2015– 2016. https://nces.ed.gov/programs/digest/d17/tables/dt17_325.50. asp?current=yes.
- Table 325.55. WEB-ONLY TABLE—Degrees in foreign languages and literatures conferred by postsecondary institutions, by level of degree and sex of student: Selected years, 1959–1960 through 2015–2016. https:// nces.ed.gov/programs/digest/d17/tables/dt17_325.55.asp?current=yes.
- Table 325.60. WEB-ONLY TABLE—Degrees in the health professions and related programs conferred by postsecondary institutions, by level of de- gree and sex of student: 1970–1971 through 2015–2016. https://nces. ed.gov/programs/digest/d17/tables/dt17_325.60.asp?current=yes.
- Table 325.65. WEB-ONLY TABLE—Degrees in mathematics and statis- tics conferred by postsecondary institutions, by level of degree and sex of student: Selected years, 1949–1950 through 2015–2016. https://nces. ed.gov/programs/digest/d17/tables/dt17_325.65.asp?current=yes.
- Table 325.70. WEB-ONLY TABLE—Degrees in the physical sciences and science technologies conferred by postsecondary institutions, by level of degree and sex of student: Selected years, 1959–1960 through 2015– 2016. https://nces.ed.gov/programs/digest/d17/tables/dt17_325.70. asp?current=yes.
- Table 325.80. WEB-ONLY TABLE—Degrees in psychology conferred by postsecondary institutions, by level of degree and sex of student: Selected years, 1949–1950 through 2015–2016. https://nces.ed.gov/programs/ digest/d17/tables/dt17_325.80.asp?current=yes.
- Table 325.85. WEB-ONLY TABLE—Degrees in public administration and social services conferred by postsecondary institutions, by level of de- gree and sex of student: 1970–1971 through 2015–2016. https://nces. ed.gov/programs/digest/d17/tables/dt17_325.85.asp?current=yes.
- Table 325.90. WEB-ONLY TABLE—Degrees in the social sciences and history conferred by postsecondary institutions, by level of degree and

sex of student: 1970–1971 through 2015–2016. https://nces.ed.gov/programs/digest/d17/tables/dt17_325.90.asp?current=yes.
  - Table 325.95. WEB-ONLY TABLE—Degrees in visual and performing arts conferred by postsecondary institutions, by level of degree and sex of student: 1970–1971 through 2015–2016. https://nces.ed.gov/programs/digest/d17/tables/dt17_325.95.asp?current=yes.
- Federal Reserve Bank of St. Louis, Economic Research, Population Total for United States. https://fred.stlouisfed.org/series/POPTOTUSA647NWDB (accessed 20 March 2018).
- Office of Management and Budget, Historical Table 3.2—Outlays by Function and Subfunction: 1962–2023. https://www.whitehouse.gov/wp-content/uploads/2018/02/hist03z2-fy2019.xlsx (accessed 20 March 2018).
- Office of Management and Budget, Historical Table 4.1—Outlays by Agency: 1962–2023. https://www.whitehouse.gov/wp-content/uploads/2018/02/hist04z1-fy2019.xlsx (accessed 25 March 2018).
- National Science Foundation, U.S. Doctorates in the 20th Century, Supplemental Table S-1: Doctoral degrees awarded, by detailed field: 1920–1999. https://wayback.archive-it.org/5902/20160210224055/http://www.nsf.gov/statistics/nsf06319/tables/tabs1.xls (accessed 23 March 2018).

METHODOLOGY

The goal of the analysis is to examine the relationship between college degree conferral rates and funding for agencies/categories related to those disciplines in the United States. Multiple linear regression models were constructed according to the following parameters:

- Degree conferral rates (including Bachelor's, Master's, and Doctoral degrees) were gathered for the 17 categories of degrees tracked by the National Center for Education Statistics' Digest of Education Statistics, ranging from 1970 to 2015. In order to control for population growth, total degrees conferred for each year and each degree area were input as a percentage of the U.S. population.
- Federal outlays ranging from 1962 to 2015 were gathered from the Office of Management and Budget Historical Tables—both by agency and by function and subfunction. In order to control for the effects of population growth, budgetary fluctuations, and inflation, these figures were input as percentages of federal outlays.

- For each discipline, models were limited to *a priori* relevant funding areas, rather than all funding areas. Thus, business degree conferral rates were compared with funding for the Small Business Administration, job training, etc. Biology and biomedical research degrees were compared with funding for the Department of Health and Human Services, the National Institutes of Health, the National Science Foundation, etc. The reason for this is that the goal of the analysis is to determine whether *a priori* relevant sources of funding can explain variation in degree conferral rates. Interest here is in *targeted* questions such as "Does funding for biomedical research explain biology degree conferral rates?" as opposed to general questions such as "What funding areas, if any, explain biology degree conferral rates?"
- For each discipline, models were made using four-, six-, and eight-year delays between funding year and degree conferral year. This is because a change in funding would not result in an immediate change in degree conferral rates. Instead, such effects would be more likely to become apparent some time later. The selection of four, six, and eight years is intended to correspond to the expected time of degree completion for bachelor's, master's, and doctoral degrees.
- For each discipline and delay period, both full and stepwise multiple linear regressions models were constructed using IBM SPSS Statistics. Only the best fit models (highest adjusted R^2 value) are recorded. In some cases the full model was the best fit; in other cases a stepwise model was the best fit. The unusually small coefficients and error estimates are due to the use of small percentage values as inputs.
- Only federal outlays were included in the analysis. Accounting for the influence of local, state, and private funding sources would involve a substantial increase in the complexity of data collection and analysis.

Agriculture Degrees

Agriculture degree conferral rates were modeled using the following funding areas as independent variables:

- Energy
- Natural Resources
- Agricultural Research
- Department of Agriculture
- Department of Energy (DOE)
- Environmental Protection Agency (EPA)

Table A.1 FULL MODEL REGRESSION RESULTS FOR AGRICULTURE DEGREES ON A FOUR-YEAR DELAY. $R^2 = 0.693$, ADJ. $R^2 = 0.645$.

Independent Variable	Coefficient	p-value
Constant	0.0002024 (0.0000172)	0.000
Energy	0.0000376 (0.0000082)	0.000
Natural Resources	0.0000566 (0.0000184)	0.004
Agricultural Research	−0.0002006 (0.0000886)	0.029
Department of Agriculture	−0.0000104 (0.0000035)	0.004
DOE	−0.0000801 (0.0000127)	0.000
EPA	−0.0000795 (0.0000251)	0.003

Table A.2 STEPWISE (BEST FIT) MODEL REGRESSION RESULTS FOR AGRICULTURE DEGREES ON A SIX-YEAR DELAY. $R^2 = 0.641$, ADJ. $R^2 = 0.616$.

Independent Variable	Coefficient	p-value
Constant	0.0001373 (0.0000084)	0.000
Department of Agriculture	−0.0000056 (0.0000022)	0.013
DOE	−0.0000363 (0.0000079)	0.000
Agricultural Research	0.0001500 (0.0000377)	0.000

Table A.3 STEPWISE (BEST FIT) MODEL REGRESSION RESULTS FOR AGRICULTURE DEGREES ON AN EIGHT-YEAR DELAY. $R^2 = 0.797$, ADJ. $R^2 = 0.777$.

Independent Variable	Coefficient	p-value
Constant	0.0001430 (0.0000067)	0.000
Department of Agriculture	−0.0000045 (0.0000021)	0.034
Agricultural Research	0.0002637 (0.0000392)	0.000
DOE	−0.0000256 (0.0000053)	0.000
Natural Resources	−0.0000278 (0.0000059)	0.000

Architecture Degrees

Architecture degree conferral rates were modeled using the following funding areas as independent variables:

- Small Business Administration
- Corps of Engineers/Civil Works
- Consumer and Occupational Health and Safety

Table A.4 FULL MODEL REGRESSION RESULTS FOR ARCHITECTURE
DEGREES ON A FOUR-YEAR DELAY. $R^2 = 0.360$, ADJ. $R^2 = 0.315$.

Independent Variable	Coefficient	p-value
Constant	0.0000441 (0.0000040)	0.000
Consumer and Occupational Health and Safety	0.0000807 (0.0000321)	0.016
Corps of Engineers/Civil Works	−0.0000122 (0.0000038)	0.002
Small Business Administration	0.0000120 (0.0000069)	0.088

Table A.5 FULL MODEL REGRESSION RESULTS FOR ARCHITECTURE
DEGREES ON A SIX-YEAR DELAY. $R^2 = 0.205$, ADJ. $R^2 = 0.148$.

Independent Variable	Coefficient	p-value
Constant	0.0000443 (0.0000047)	0.000
Consumer and Occupational Health and Safety	0.0000736 (0.0000360)	0.047
Corps of Engineers/Civil Works	−0.0000066 (0.0000037)	0.081
Small Business Administration	0.0000021 (0.0000080)	0.797

Table A.6 STEPWISE (BEST FIT) MODEL REGRESSION RESULTS
FOR ARCHITECTURE DEGREES ON AN EIGHT-YEAR DELAY.
$R^2 = 0.116$, ADJ. $R^2 = 0.096$.

Independent Variable	Coefficient	p-value
Constant	0.0000432 (0.0000036)	0.000
Consumer and Occupational Health and Safety	0.0000636 (0.0000264)	0.020

Biology and Biomedical Research Degrees

Biology and Biomedical Research degree conferral rates were modeled using
the following funding areas as independent variables:

- National Institutes of Health (NIH)
- National Aeronautics and Space Administration (NASA)
- EPA
- National Science Foundation (NSF)
- Health Research and Training
- Healthcare Services
- Space Science
- Department of Health and Human Services (HHS)
- General Science

Table A.7 FULL MODEL REGRESSION RESULTS FOR BIOLOGY AND BIOMEDICAL RESEARCH DEGREES ON A FOUR-YEAR DELAY. $R^2 = 0.909$, ADJ. $R^2 = 0.895$.

Independent Variable	Coefficient	p-value
Constant	0.0001311 (0.0000262)	0.000
Healthcare Services	0.0000248 (0.0000036)	0.000
Health Research and Training	0.0003925 (0.0000584)	0.000
EPA	−0.0000887 (0.0000197)	0.000
NASA	0.0000252 (0.0000117)	0.037
NSF	−0.0003662 (0.0002526)	0.155
NIH	−0.0002992 (0.0000434)	0.000

Table A.8 FULL MODEL REGRESSION RESULTS FOR BIOLOGY AND BIOMEDICAL RESEARCH DEGREES ON A SIX-YEAR DELAY. $R^2 = 0.885$, ADJ. $R^2 = 0.867$.

Independent Variable	Coefficient	p-value
Constant	0.0002049 (0.0000296)	0.000
Healthcare Services	0.0000148 (0.0000048)	0.004
Health Research and Training	0.0003813 (0.0000689)	0.000
EPA	−0.0001526 (0.0000220)	0.000
NASA	0.0000266 (0.0000135)	0.056
NSF	−0.0008583 (0.0002840)	0.004
NIH	−0.0001472 (0.0000569)	0.014

Table A.9 FULL MODEL REGRESSION RESULTS FOR BIOLOGY AND BIOMEDICAL RESEARCH DEGREES ON AN EIGHT-YEAR DELAY. $R^2 = 0.866$, ADJ. $R^2 = 0.846$.

Independent Variable	Coefficient	p-value
Constant	0.0001979 (0.0000288)	0.000
Healthcare Services	0.0000124 (0.0000035)	0.001
Health Research and Training	0.0002084 (0.0000698)	0.005
EPA	−0.0001298 (0.0000220)	0.000
NASA	0.0000228 (0.0000116)	0.058
NSF	−0.0006598 (0.0002968)	0.032
NIH	0.0000127 (0.0000466)	0.786

There were three cases of multicollinearity: funding for space science and for NASA; funding for HHS and healthcare services; and funding for general science and the NSF. The former of each pair was excluded from the models below.

Business Degrees

Business degree conferral rates were modeled using the following funding areas as independent variables:

- Small Business Administration
- Other Labor Services
- Social Services
- Job Training and Employment

Table A.10 STEPWISE (BEST FIT) MODEL REGRESSION RESULTS FOR BUSINESS DEGREES ON A FOUR-YEAR DELAY. R^2 = 0.396, ADJ. R^2 = 0.382.

Independent Variable	Coefficient	p-value
Constant	0.0018724 (0.0001109)	0.000
Social Services	−0.0006617 (0.0001232)	0.000

Table A.11 STEPWISE (BEST FIT) MODEL REGRESSION RESULTS FOR BUSINESS DEGREES ON A SIX-YEAR DELAY. R^2 = 0.113, ADJ. R^2 = 0.092.

Independent Variable	Coefficient	p-value
Constant	0.0014538 (0.0000751)	0.000
Job Training and Employment	−0.0001976 (0.0000837)	0.023

Table A.12 FULL MODEL REGRESSION RESULTS FOR BUSINESS DEGREES ON AN EIGHT-YEAR DELAY. R^2 = 0.160, ADJ. R^2 = 0.078.

Independent Variable	Coefficient	p-value
Constant	0.0013946 (0.0006032)	0.026
Job Training and Employment	−0.0000706 (0.0002309)	0.761
Other Labor Services	−0.0028193 (0.0092436)	0.762
Social Services	0.0003158 (0.0002098)	0.140
Small Business Administration	−0.0007311 (0.0005366)	0.180

Communications Degrees

Communications degree conferral rates were modeled using the following funding areas as independent variables:

- Small Business Administration
- Healthcare Services
- Social Services
- Other Labor Services
- Job Training and Employment

Table A.13 STEPWISE (BEST FIT) MODEL REGRESSION RESULTS FOR COMMUNICATIONS DEGREES ON A FOUR-YEAR DELAY. $R^2 = 0.863$, ADJ. $R^2 = 0.853$.

Independent Variable	Coefficient	p-value
Constant	0.0000645 (0.0000423)	0.135
Healthcare Services	0.0000230 (0.0000021)	0.000
Social Services	−0.0001004 (0.0000214)	0.000
Other Labor Services	0.0017439 (0.0005964)	0.006

Table A.14 FULL MODEL REGRESSION RESULTS FOR COMMUNICATIONS DEGREES ON A SIX-YEAR DELAY. $R^2 = 0.863$, ADJ. $R^2 = 0.846$.

Independent Variable	Coefficient	p-value
Constant	0.0000557 (0.0000559)	0.325
Job Training and Employment	0.0000368 (0.0000230)	0.118
Other Labor Services	0.0007461 (0.0008402)	0.380
Social Services	−0.0000408 (0.0000219)	0.069
Healthcare Services	0.0000273 (0.0000020)	0.000
Small Business Administration	−0.0000996 (0.0000563)	0.084

Table A.15 FULL MODEL REGRESSION RESULTS FOR COMMUNICATIONS DEGREES ON AN EIGHT-YEAR DELAY. $R^2 = 0.912$, ADJ. $R^2 = 0.901$.

Independent Variable	Coefficient	p-value
Constant	0.0000231 (0.0000506)	0.650
Job Training and Employment	0.0000307 (0.0000186)	0.107
Other Labor Services	0.0010068 (0.0007539)	0.189
Social Services	−0.0000113 (0.0000172)	0.513
Healthcare Services	0.0000281 (0.0000015)	0.000
Small Business Administration	−0.0001040 (0.0000436)	0.022

Computer Science Degrees

Computer Science degree conferral rates were modeled using the following funding areas as independent variables:

- NIH
- NASA
- Health Research and Training
- Military Research and Development
- Corps of Engineers/Civil Works
- NSF
- Space Science
- General Science

There were two cases of multicollinearity: funding for space science and for NASA; and funding for general science and the NSF. The former of each pair was excluded from the models below.

Table A.16 STEPWISE (BEST FIT) MODEL REGRESSION RESULTS FOR COMPUTER SCIENCE DEGREES ON A FOUR-YEAR DELAY. $R^2 = 0.790$, ADJ. $R^2 = 0.775$.

Independent Variable	Coefficient	p-value
Constant	0.0004513 (0.0000340)	0.000
Military Research and Development	−0.0000870 (0.0000162)	0.000
Corps of Engineers/Civil Works	−0.0002482 (0.0000450)	0.000
NASA	0.0000411 (0.0000169)	0.019

Table A.17 STEPWISE (BEST FIT) MODEL REGRESSION RESULTS FOR COMPUTER SCIENCE DEGREES ON A SIX-YEAR DELAY. $R^2 = 0.804$, ADJ. $R^2 = 0.785$.

Independent Variable	Coefficient	p-value
Constant	0.0003298 (0.0000369)	0.000
Corps of Engineers/Civil Works	−0.0000864 (0.0000431)	0.052
Military Research and Development	−0.0000527 (0.0000103)	0.000
NIH	0.0001691 (0.0000495)	0.001
Health Research and Training	−0.0001323 (0.0000595)	0.032

Table A.18 STEPWISE (BEST FIT) MODEL REGRESSION RESULTS FOR COMPUTER SCIENCE DEGREES ON AN EIGHT-YEAR DELAY. $R^2 = 0.804$, ADJ. $R^2 = 0.790$.

Independent Variable	Coefficient	p-value
Constant	0.0002490 (0.0000281)	0.000
Corps of Engineers/Civil Works	−0.0001069 (0.0000394)	0.010
Military Research and Development	−0.0000390 (0.0000088)	0.000
NIH	0.0001045 (0.0000286)	0.001

Education Degrees

Education degree conferral rates were modeled using the following funding areas as independent variables:

- Department of Education
- Education Research
- Elementary, Secondary, and Vocational Education
- Higher Education

Table A.19 FULL MODEL REGRESSION RESULTS FOR EDUCATION DEGREES ON A FOUR-YEAR DELAY. $R^2 = 0.404$, ADJ. $R^2 = 0.344$.

Independent Variable	Coefficient	p-value
Constant	0.0006161 (0.0001708)	0.001
Elementary, Secondary, and Vocational Education	−0.0020774 (0.0006758)	0.004
Higher Education	−0.0025619 (0.0007162)	0.001
Education Research	−0.0030901 (0.0010285)	0.005
Department of Education	0.0024805 (0.0007041)	0.001

Table A.20 FULL MODEL REGRESSION RESULTS FOR EDUCATION DEGREES ON A SIX-YEAR DELAY. $R^2 = 0.456$, ADJ. $R^2 = 0.403$.

Independent Variable	Coefficient	p-value
Constant	0.0011832 (0.0001286)	0.000
Elementary, Secondary, and Vocational Education	0.0018384 (0.0009043)	0.049
Higher Education	0.0013510 (0.0008836)	0.134
Education Research	−0.0008789 (0.0011331)	0.442
Department of Education	−0.0015726 (0.0008859)	0.083

Table A.21 FULL MODEL REGRESSION RESULTS FOR EDUCATION
DEGREES ON AN EIGHT-YEAR DELAY. $R^2 = 0.826$, ADJ. $R^2 = 0.809$.

Independent Variable	Coefficient	p-value
Constant	0.0014600 (0.0000629)	0.000
Elementary, Secondary, and Vocational Education	0.0023325 (0.0005037)	0.000
Higher Education	0.0017515 (0.0004890)	0.001
Education Research	−0.0010115 (0.0006449)	0.125
Department of Education	−0.0021206 (0.0004892)	0.000

Engineering Degrees

Engineering degree conferral rates were modeled using the following funding areas as independent variables:

- NIH
- NASA
- Natural Resources
- Department of Agriculture
- Health Research and Training
- Energy
- DOE
- Military Research and Development
- NSF
- Agricultural Research
- EPA
- HHS
- Corps of Engineers/Civil Works
- Space Science
- General Science

There were two cases of multicollinearity: funding for space science and for NASA; and funding for general science and the NSF. The former of each pair was excluded from the models below.

Table A.22 FULL MODEL REGRESSION RESULTS FOR ENGINEERING DEGREES ON A FOUR-YEAR DELAY. $R^2 = 0.932$, ADJ. $R^2 = 0.904$.

Independent Variable	Coefficient	p-value
Constant	0.0002869 (0.0001349)	0.041
Military Research and Development	0.0000692 (0.0000212)	0.003
Energy	0.0000887 (0.0000287)	0.004
Natural Resources	−0.0001444 (0.0000715)	0.052
Agricultural Research	−0.0010678 (0.0003796)	0.008
Health Research and Training	0.0001271 (0.0000754)	0.101
Department of Agriculture	0.0000111 (0.0000088)	0.216
DOE	−0.0000706 (0.0000507)	0.173
HHS	0.0000163 (0.0000051)	0.003
Corps of Engineers/Civil Works	0.0006467 (0.0001347)	0.000
EPA	0.0001543 (0.0000928)	0.106
NASA	0.0000253 (0.0000261)	0.339
NSF	−0.0019009 (0.0004731)	0.000
NIH	0.0000467 (0.0000617)	0.454

Table A.23 FULL MODEL REGRESSION RESULTS FOR ENGINEERING DEGREES ON A SIX-YEAR DELAY. $R^2 = 0.892$, ADJ. $R^2 = 0.849$.

Independent Variable	Coefficient	p-value
Constant	0.0001580 (0.0001379)	0.260
Military Research and Development	0.0000862 (0.0000239)	0.001
Energy	0.0001111 (0.0000302)	0.001
Natural Resources	−0.0001955 (0.0000898)	0.037
Agricultural Research	−0.0006667 (0.0004632)	0.160
Health Research and Training	0.0000671 (0.0000985)	0.500
Department of Agriculture	0.0000108 (0.0000110)	0.331
DOE	−0.0001768 (0.0000487)	0.001
HHS	0.0000212 (0.0000062)	0.002
Corps of Engineers/Civil Works	0.0006930 (0.0001814)	0.001
EPA	0.0002984 (0.0001158)	0.015
NASA	0.0000337 (0.0000275)	0.230
NSF	−0.0016040 (0.0005544)	0.007
NIH	0.0000700 (0.0000852)	0.418

Table A.24 FULL MODEL REGRESSION RESULTS FOR ENGINEERING
DEGREES ON AN EIGHT-YEAR DELAY. $R^2 = 0.876$, ADJ. $R^2 = 0.826$.

Independent Variable	Coefficient	p-value
Constant	0.0000153 (0.0001482)	0.918
Military Research and Development	0.0000621 (0.0000257)	0.022
Energy	0.0000549 (0.0000313)	0.089
Natural Resources	−0.0001674 (0.0000980)	0.097
Agricultural Research	0.0001318 (0.0005163)	0.800
Health Research and Training	0.0000864 (0.0000982)	0.386
Department of Agriculture	0.0000137 (0.0000097)	0.166
DOE	−0.0001117 (0.0000478)	0.026
HHS	0.0000148 (0.0000069)	0.039
Corps of Engineers/Civil Works	0.0004130 (0.0002045)	0.052
EPA	0.0003131 (0.0001212)	0.015
NASA	−0.0000088 (0.0000187)	0.641
NSF	−0.0006257 (0.0005751)	0.285
NIH	0.0000649 (0.0000881)	0.467

English and Literature Degrees

English and Literature degree conferral rates were modeled using the following
funding areas as independent variables:

- National Endowment for the Humanities (NEH)
- Department of Education
- Higher Education
- Education Research
- Elementary, Secondary, and Vocational Education

Table A.25 STEPWISE (BEST FIT) MODEL REGRESSION RESULTS FOR
ENGLISH AND LITERATURE DEGREES ON A FOUR-YEAR DELAY.
$R^2 = 0.318$, ADJ. $R^2 = 0.302$.

Independent Variable	Coefficient	p-value
Constant	0.0002620 (0.0000114)	0.000
NEH	−0.0038445 (0.0008495)	0.000

Table A.26 FULL MODEL REGRESSION RESULTS FOR ENGLISH AND
LITERATURE DEGREES ON A SIX-YEAR DELAY. $R^2 = 0.719$, ADJ. $R^2 = 0.684$.

Independent Variable	Coefficient	p-value
Constant	0.0003701 (0.0000226)	0.000
Elementary, Secondary, and Vocational Education	0.0005256 (0.0001522)	0.001
Higher Education	0.0005668 (0.0001494)	0.000
Education Research	0.0003146 (0.0002851)	0.276
Department of Education	−0.0005839 (0.0001495)	0.000
NEH	−0.0027671 (0.0012549)	0.033

Table A.27 STEPWISE (BEST FIT) MODEL REGRESSION RESULTS FOR
ENGLISH AND LITERATURE DEGREES ON AN EIGHT-YEAR DELAY.
$R^2 = 0.897$, ADJ. $R^2 = 0.890$.

Independent Variable	Coefficient	p-value
Constant	0.0003825 (0.0000098)	0.000
Department of Education	−0.0005155 (0.0000452)	0.000
Higher Education	0.0004839 (0.0000468)	0.000
Elementary, Secondary, and Vocational Education	0.0004562 (0.0000494)	0.000

Foreign Languages Degrees

Foreign Languages degree conferral rates were modeled using the following
funding areas as independent variables:

- NEH
- Department of Education
- Higher Education
- Education Research
- Elementary, Secondary, and Vocational Education

Table A.28 FULL MODEL REGRESSION RESULTS FOR FOREIGN
LANGUAGES DEGREES ON A FOUR-YEAR DELAY. $R^2 = 0.451$, ADJ. $R^2 = 0.382$.

Independent Variable	Coefficient	p-value
Constant	0.0000713 (0.0000147)	0.000
Elementary, Secondary, and Vocational Education	−0.0000530 (0.0000607)	0.388
Higher Education	−0.0000699 (0.0000661)	0.297
Education Research	−0.0000108 (0.0001535)	0.944
Department of Education	0.0000693 (0.0000648)	0.291
NEH	−0.0015746 (0.0006715)	0.024

Table A.29 FULL MODEL REGRESSION RESULTS FOR FOREIGN LANGUAGES DEGREES ON A SIX-YEAR DELAY. $R^2 = 0.703$, ADJ. $R^2 = 0.666$.

Independent Variable	Coefficient	p-value
Constant	0.0001078 (0.0000088)	0.000
Elementary, Secondary, and Vocational Education	0.0002476 (0.0000594)	0.000
Higher Education	0.0002344 (0.0000583)	0.000
Education Research	0.0002083 (0.0001112)	0.068
Department of Education	−0.0002437 (0.0000583)	0.000
NEH	−0.0015739 (0.0004896)	0.003

Table A.30 FULL MODEL REGRESSION RESULTS FOR FOREIGN LANGUAGES DEGREES ON AN EIGHT-YEAR DELAY. $R^2 = 0.877$, ADJ. $R^2 = 0.862$.

Independent Variable	Coefficient	p-value
Constant	0.0001263 (0.0000052)	0.000
Elementary, Secondary, and Vocational Education	0.0002198 (0.0000377)	0.000
Higher Education	0.0002031 (0.0000367)	0.000
Education Research	0.0000865 (0.0000752)	0.257
Department of Education	−0.0002211 (0.0000366)	0.000
NEH	−0.0009062 (0.0003374)	0.010

Health Professions Degrees

Health Professions degree conferral rates were modeled using the following funding areas as independent variables:

- NIH
- Consumer and Occupational Health and Safety
- Health Research and Training
- Healthcare Services
- HHS

There was one case of multicollinearity: funding for HHS and healthcare services. The former was excluded from the models below.

Table A.31 FULL MODEL REGRESSION RESULTS FOR HEALTH PROFESSIONS DEGREES ON A FOUR YEAR DELAY. $R^2 = 0.809$, ADJ. $R^2 = 0.791$.

Independent Variable	Coefficient	p-value
Constant	0.0003153 (0.0001231)	0.014
Healthcare Services	0.0001077 (0.0000117)	0.000
Health Research and Training	0.0007107 (0.0001821)	0.000
Consumer and Occupational Health and Safety	−0.0010842 (0.0007226)	0.141
NIH	−0.0009423 (0.0002019)	0.000

Table A.32 FULL MODEL REGRESSION RESULTS FOR HEALTH PROFESSIONS DEGREES ON A SIX-YEAR DELAY. $R^2 = 0.780$, ADJ. $R^2 = 0.759$.

Independent Variable	Coefficient	p-value
Constant	0.0002667 (0.0001327)	0.051
Healthcare Services	0.0000796 (0.0000113)	0.000
Health Research and Training	0.0006567 (0.0001963)	0.002
Consumer and Occupational Health and Safety	−0.0017203 (0.0007857)	0.034
NIH	−0.0004717 (0.0002035)	0.026

Table A.33 FULL MODEL REGRESSION RESULTS FOR HEALTH PROFESSIONS DEGREES ON AN EIGHT-YEAR DELAY. $R^2 = 0.844$, ADJ. $R^2 = 0.829$.

Independent Variable	Coefficient	p-value
Constant	0.0001056 (0.0001122)	0.352
Healthcare Services	0.0000624 (0.0000089)	0.000
Health Research and Training	0.0005593 (0.0001606)	0.001
Consumer and Occupational Health and Safety	−0.0013140 (0.0006893)	0.064
NIH	−0.0000782 (0.0001602)	0.628

Mathematics and Statistics Degrees

Mathematics and Statistics degree conferral rates were modeled using the following funding areas as independent variables:

- NSF
- DOE
- Military Research and Development

- Energy
- NASA
- Space Science
- General Science

There were two cases of multicollinearity: funding for space science and for NASA; and funding for general science and the NSF. The former of each pair was excluded from the models below.

Table A.34 FULL MODEL REGRESSION RESULTS FOR MATHEMATICS AND STATISTICS DEGREES ON A FOUR-YEAR DELAY. R^2 = 0.649, ADJ. R^2 = 0.606.

Independent Variable	Coefficient	p-value
Constant	0.0000223 (0.0000327)	0.500
Military Research and Development	0.0000116 (0.0000063)	0.074
Energy	0.0000053 (0.0000069)	0.449
DOE	−0.0000060 (0.0000217)	0.783
NASA	0.0000132 (0.0000087)	0.138
NSF	0.0000811 (0.0000973)	0.410

Table A.35 STEPWISE (BEST FIT) MODEL REGRESSION RESULTS FOR MATHEMATICS AND STATISTICS DEGREES ON A SIX-YEAR DELAY. R^2 = 0.624, ADJ. R^2 = 0.616.

Independent Variable	Coefficient	p-value
Constant	0.0000572 (0.0000034)	0.000
NASA	0.0000190 (0.0000022)	0.000

Table A.36 STEPWISE (BEST FIT) MODEL REGRESSION RESULTS FOR MATHEMATICS AND STATISTICS DEGREES ON AN EIGHT-YEAR DELAY. R^2 = 0.721, ADJ. R^2 = 0.693.

Independent Variable	Coefficient	p-value
Constant	−0.0000193 (0.0000162)	0.239
DOE	0.0000556 (0.0000069)	0.000
Energy	−0.0000100 (0.0000038)	0.013
NSF	0.0002639 (0.0000772)	0.001
NASA	−0.0000105 (0.0000040)	0.013

Physical Sciences Degrees

Physical Sciences degree conferral rates were modeled using the following funding areas as independent variables:

- NSF
- DOE
- EPA
- Military Research and Development
- Energy
- NASA
- Natural Resources
- Space Science
- General Science

There were two cases of multicollinearity: funding for space science and for NASA; and funding for general science and the NSF. The former of each pair was excluded from the models below.

Table A.37 FULL MODEL REGRESSION RESULTS FOR PHYSICAL SCIENCES DEGREES ON A FOUR-YEAR DELAY. $R^2 = 0.857$, ADJ. $R^2 = 0.830$.

Independent Variable	Coefficient	p-value
Constant	0.0001371 (0.0000205)	0.000
Military Research and Development	0.0000038 (0.0000034)	0.274
Energy	0.0000484 (0.0000064)	0.000
EPA	−0.0000223 (0.0000238)	0.355
Natural Resources	−0.0000012 (0.0000155)	0.937
DOE	−0.0000785 (0.0000127)	0.000
NASA	0.0000166 (0.0000062)	0.011
NSF	0.0001423 (0.0000686)	0.045

Table A.38 FULL MODEL REGRESSION RESULTS FOR PHYSICAL SCIENCES DEGREES ON A SIX-YEAR DELAY. $R^2 = 0.801$, ADJ. $R^2 = 0.764$.

Independent Variable	Coefficient	p-value
Constant	0.0000503 (0.0000165)	0.004
Military Research and Development	0.0000157 (0.0000036)	0.000
Energy	0.0000191 (0.0000057)	0.002
Natural Resources	0.0000198 (0.0000169)	0.249
DOE	−0.0000358 (0.0000096)	0.001
EPA	−0.0000213 (0.0000280)	0.452
NASA	−0.0000038 (0.0000057)	0.506
NSF	0.0001945 (0.0000804)	0.020

Table A.39 FULL MODEL REGRESSION RESULTS FOR PHYSICAL SCIENCES
DEGREES ON AN EIGHT-YEAR DELAY. $R^2 = 0.877$, ADJ. $R^2 = 0.854$.

Independent Variable	Coefficient	p-value
Constant	0.0000174 (0.0000131)	0.190
Military Research and Development	0.0000171 (0.0000027)	0.000
Energy	0.0000080 (0.0000043)	0.067
Natural Resources	0.0000204 (0.0000100)	0.048
DOE	−0.0000158 (0.0000071)	0.032
EPA	−0.0000187 (0.0000183)	0.314
NASA	−0.0000139 (0.0000024)	0.000
NSF	0.0003008 (0.0000593)	0.000

Psychology Degrees

Psychology degree conferral rates were modeled using the following funding
areas as independent variables:

- NIH
- Consumer and Occupational Health and Safety
- NSF
- Health Research and Training
- Healthcare Services
- Social Services
- HHS
- General Science

There were two cases of multicollinearity: funding for HHS and for healthcare
services; and funding for general science and the NSF. The former of each pair
was excluded from the models below.

Table A.40 FULL MODEL REGRESSION RESULTS FOR PSYCHOLOGY
DEGREES ON A FOUR-YEAR DELAY. $R^2 = 0.936$, ADJ. $R^2 = 0.926$.

Independent Variable	Coefficient	p-value
Constant	0.0003748 (0.0000325)	0.000
Social Services	0.0000201 (0.0000320)	0.534
Healthcare Services	0.0000220 (0.0000032)	0.000
Health Research and Training	0.0002307 (0.0000503)	0.000
Consumer and Occupational Health and Safety	−0.0014918 (0.0002641)	0.000
NSF	−0.0003466 (0.0001810)	0.063
NIH	−0.0001332 (0.0000439)	0.004

Table A.41 FULL MODEL REGRESSION RESULTS FOR PSYCHOLOGY DEGREES
ON A SIX-YEAR DELAY. $R^2 = 0.942$, ADJ. $R^2 = 0.933$.

Independent Variable	Coefficient	p-value
Constant	0.0004410 (0.0000291)	0.000
Social Services	0.0000267 (0.0000222)	0.237
Healthcare Services	0.0000131 (0.0000033)	0.000
Health Research and Training	0.0001382 (0.0000479)	0.006
Consumer and Occupational Health and Safety	−0.0017221 (0.0002099)	0.000
NSF	−0.0005957 (0.0001636)	0.001
NIH	0.0000472 (0.0000516)	0.366

Table A.42 FULL MODEL REGRESSION RESULTS FOR PSYCHOLOGY DEGREES
ON AN EIGHT-YEAR DELAY. $R^2 = 0.907$, ADJ. $R^2 = 0.893$.

Independent Variable	Coefficient	p-value
Constant	0.0003614 (0.0000347)	0.000
Social Services	−0.0000127 (0.0000285)	0.659
Healthcare Services	0.0000203 (0.0000035)	0.000
Health Research and Training	0.0000256 (0.0000622)	0.683
Consumer and Occupational Health and Safety	−0.0009472 (0.0002625)	0.001
NSF	−0.0002200 (0.0001886)	0.251
NIH	0.0000487 (0.0000619)	0.436

Public Administration and Social Work Degrees

Public Administration and Social Work degree conferral rates were modeled
using the following funding areas as independent variables:

- Department of Education
- Education Research
- Healthcare Services
- Elementary, Secondary, and Vocational Education
- Social Services
- Other Labor Services
- Health Research and Training
- Job Training and Employment
- Consumer and Occupational Health and Safety
- Higher Education

Table A.43 FULL MODEL REGRESSION RESULTS FOR PUBLIC
ADMINISTRATION AND SOCIAL WORK DEGREES ON A FOUR-YEAR DELAY.
$R^2 = 0.869$, ADJ. $R^2 = 0.832$.

Independent Variable	Coefficient	p-value
Constant	0.0001595 (0.0000380)	0.000
Elementary, Secondary, and Vocational Education	0.0000799 (0.0001096)	0.470
Higher Education	0.0000821 (0.0001305)	0.533
Education Research	−0.0000918 (0.0002044)	0.656
Job Training and Employment	0.0000053 (0.0000138)	0.703
Other Labor Services	−0.0026798 (0.0006940)	0.000
Social Services	−0.0000084 (0.0000307)	0.786
Healthcare Services	0.0000076 (0.0000026)	0.006
Health Research and Training	0.0000446 (0.0000457)	0.336
Consumer and Occupational Health and Safety	0.0010721 (0.0003373)	0.003
Department of Education	−0.0000815 (0.0001299)	0.535

Table A.44 FULL MODEL REGRESSION RESULTS FOR PUBLIC
ADMINISTRATION AND SOCIAL WORK DEGREES ON A SIX-YEAR DELAY.
$R^2 = 0.876$, ADJ. $R^2 = 0.840$.

Independent Variable	Coefficient	p-value
Constant	0.0001364 (0.0000345)	0.000
Elementary, Secondary, and Vocational Education	−0.0001939 (0.0001506)	0.207
Higher Education	−0.0002508 (0.0001624)	0.131
Education Research	−0.0001349 (0.0002079)	0.521
Job Training and Employment	−0.0000088 (0.0000132)	0.512
Other Labor Services	−0.0014098 (0.0006134)	0.028
Social Services	−0.0000357 (0.0000319)	0.271
Healthcare Services	0.0000069 (0.0000022)	0.004
Health Research and Training	−0.0000310 (0.0000349)	0.380
Consumer and Occupational Health and Safety	0.0002655 (0.0003280)	0.424
Department of Education	0.0002677 (0.0001628)	0.109

Table A.45 STEPWISE (BEST FIT) MODEL REGRESSION RESULTS FOR PUBLIC ADMINISTRATION AND SOCIAL WORK DEGREES ON AN EIGHT-YEAR DELAY. $R^2 = 0.903$, ADJ. $R^2 = 0.896$.

Independent Variable	Coefficient	p-value
Constant	0.0000642 (0.0000084)	0.000
Healthcare Services	0.0000077 (0.0000011)	0.000
Department of Education	0.0000427 (0.0000056)	0.000
Job Training and Employment	−0.0000328 (0.0000051)	0.000

Social Sciences and History Degrees

Social Sciences and History degree conferral rates were modeled using the following funding areas as independent variables:

- NEH
- Department of Education
- Social Services
- NIH
- Elementary, Secondary, and Vocational Education
- Other Labor Services
- Education Research
- NSF
- Job Training and Employment
- Higher Education
- General Science

There was one case of multicollinearity: funding for general science and the NSF. The former was excluded from the models below.

Table A.46 STEPWISE (BEST FIT) MODEL REGRESSION RESULTS FOR SOCIAL SCIENCES AND HISTORY DEGREES ON A FOUR-YEAR DELAY. $R^2 = 0.847$, ADJ. $R^2 = 0.832$.

Independent Variable	Coefficient	p-value
Constant	0.0004016 (0.0000542)	0.000
NSF	0.0025916 (0.0002834)	0.000
Education Research	−0.0008237 (0.0003054)	0.010
NIH	−0.0001242 (0.0000402)	0.004
Social Services	−0.0001120 (0.0000490)	0.027

Table A.47 FULL MODEL REGRESSION RESULTS FOR SOCIAL SCIENCES AND HISTORY DEGREES ON A SIX YEAR DELAY. $R^2 = 0.928$, ADJ. $R^2 = 0.907$.

Independent Variable	Coefficient	p-value
Constant	0.0004098 (0.0000703)	0.000
Elementary, Secondary, and Vocational Education	0.0005224 (0.0002726)	0.064
Higher Education	0.0005246 (0.0002919)	0.081
Education Research	−0.0001697 (0.0003711)	0.650
Job Training and Employment	−0.0000297 (0.0000268)	0.275
Other Labor Services	0.0023481 (0.0011355)	0.046
Social Services	−0.0000864 (0.0000616)	0.169
Department of Education	−0.0005283 (0.0002921)	0.079
NSF	0.0014337 (0.0002804)	0.000
NIH	−0.0000272 (0.0000489)	0.582
NEH	−0.0033270 (0.0025797)	0.206

Table A.48 FULL MODEL REGRESSION RESULTS FOR SOCIAL SCIENCES AND HISTORY DEGREES ON AN EIGHT-YEAR DELAY. $R^2 = 0.931$, ADJ. $R^2 = 0.912$.

Independent Variable	Coefficient	p-value
Constant	0.0005124 (0.0000826)	0.000
Elementary, Secondary, and Vocational Education	0.0005207 (0.0002652)	0.058
Higher Education	0.0005803 (0.0002945)	0.057
Education Research	0.0000834 (0.0003809)	0.828
Job Training and Employment	−0.0000126 (0.0000272)	0.645
Other Labor Services	0.0033326 (0.0010852)	0.004
Social Services	−0.0000361 (0.0000711)	0.615
Department of Education	−0.0006041 (0.0002890)	0.044
NSF	0.0003047 (0.0002612)	0.251
NIH	0.0001069 (0.0000505)	0.041
NEH	−0.0067686 (0.0024767)	0.010

Visual and Performing Arts Degrees

Visual and Performing Arts degree conferral rates were modeled using the following funding areas as independent variables:

- NEH
- Department of Education

- Higher Education
- Education Research
- Elementary, Secondary, and Vocational Education

Table A.49 STEPWISE (BEST FIT) MODEL REGRESSION RESULTS FOR
VISUAL AND PERFORMING ARTS DEGREES ON A FOUR-YEAR DELAY.
$R^2 = 0.476$, ADJ. $R^2 = 0.451$.

Independent Variable	Coefficient	p-value
Constant	0.0002219 (0.0000351)	0.000
NEH	−0.0040477 (0.0010479)	0.000
Elementary, Secondary, and Vocational Education	0.0000706 (0.0000235)	0.004

Table A.50 FULL MODEL REGRESSION RESULTS FOR VISUAL AND
PERFORMING ARTS DEGREES ON A SIX-YEAR DELAY. $R^2 = 0.562$, ADJ.
$R^2 = 0.508$.

Independent Variable	Coefficient	p-value
Constant	0.0001898 (0.0000362)	0.000
Elementary, Secondary, and Vocational Education	−0.0005719 (0.0002430)	0.024
Higher Education	−0.0005556 (0.0002386)	0.025
Education Research	−0.0003133 (0.0004553)	0.495
Department of Education	0.0006141 (0.0002387)	0.014
NEH	−0.0063089 (0.0020038)	0.003

Table A.51 FULL MODEL REGRESSION RESULTS FOR VISUAL AND
PERFORMING ARTS DEGREES ON AN EIGHT-YEAR DELAY.
$R^2 = 0.636$, ADJ. $R^2 = 0.590$.

Independent Variable	Coefficient	p-value
Constant	0.0001787 (0.0000299)	0.000
Elementary, Secondary, and Vocational Education	−0.0007431 (0.0002183)	0.002
Higher Education	−0.0005936 (0.0002129)	0.008
Education Research	0.0002052 (0.0004362)	0.641
Department of Education	0.0007089 (0.0002122)	0.002
NEH	−0.0096396 (0.0019560)	0.000

PHD CONFERRAL RATES

Table A.52 CORRELATIONS BETWEEN PHD CONFERRAL RATES (ACTUAL) BY DISCIPLINE WITH NASA'S BUDGET (IN ACTUAL DOLLARS) AND BY TOTAL FEDERAL OUTLAYS (IN ACTUAL DOLLARS), FROM 1960 TO 1999, DURING THE APOLLO ERA (1960–1975), AND AFTER THE APOLLO ERA (1976–1999), FOR DELAYS OF FOUR, SIX, AND EIGHT YEARS FROM FUNDING YEAR TO DEGREE CONFERRAL YEAR.

	1960–1999 NASA (4 year)	1960–1999 Total (4 year)	1960–1999 NASA (6 year)	1960–1999 Total (6 year)	1960–1999 NASA (8 year)	1960–1999 Total (8 year)	Apollo Era (1960–1975) NASA (4 year)	Apollo Era (1960–1975) Total (4 year)	Apollo Era (1960–1975) NASA (6 year)	Apollo Era (1960–1975) Total (6 year)	Apollo Era (1960–1975) NASA (8 year)	Apollo Era (1960–1975) Total (8 year)	Post-Apollo (1976–1999) NASA (4 year)	Post-Apollo (1976–1999) Total (4 year)	Post-Apollo (1976–1999) NASA (6 year)	Post-Apollo (1976–1999) Total (6 year)	Post-Apollo (1976–1999) NASA (8 year)	Post-Apollo (1976–1999) Total (8 year)
All Fields	0.855	0.811	0.906	0.841	0.912	0.892	0.024	0.961	0.127	0.969	-0.168	0.839	0.973	0.961	0.916	0.965	0.842	0.958
Agricultural and biological sciences	0.657	0.743	0.611	0.705	0.506	0.678	0.537	0.529	0.319	0.291	0.567	0.811	0.109	0.239	-0.091	0.001	-0.255	-0.174
Biological sciences	0.939	0.911	0.952	0.934	0.905	0.962	0.020	0.942	0.050	0.973	0.308	0.927	0.973	0.946	0.954	0.964	0.900	0.969
Earth, atmospheric, and ocean sciences	0.817	0.841	0.849	0.860	0.866	0.871	-0.120	0.842	0.091	0.852	-0.218	0.624	0.847	0.872	0.814	0.885	0.766	0.882
Mathematics and computer sciences	0.950	0.883	0.905	0.892	0.812	0.899	0.064	0.956	0.208	0.938	-0.056	0.920	0.934	0.946	0.841	0.935	0.755	0.912
Physical sciences	0.774	0.580	0.671	0.563	0.459	0.561	0.302	0.974	0.415	0.911	-0.031	0.937	0.890	0.930	0.773	0.894	0.656	0.839
Psychology	0.668	0.754	0.674	0.733	0.709	0.732	-0.026	0.516	-0.427	0.662	0.748	0.452	0.803	0.767	0.898	0.840	0.944	0.932
Social sciences	0.636	0.582	0.707	0.554	0.783	0.551	-0.354	0.702	-0.172	0.882	-0.756	-0.432	0.927	0.866	0.933	0.921	0.897	0.956
Engineering	0.957	0.903	0.897	0.911	0.785	0.913	0.208	0.971	0.332	0.961	-0.007	0.954	0.932	0.956	0.841	0.935	0.734	0.896
Education	0.323	0.313	0.287	0.168	0.200	-0.056	-0.689	-0.638	-0.912	-0.385	-0.191	-0.889	-0.467	-0.620	-0.287	-0.406	0.049	0.074
Health sciences	0.945	0.977	0.931	0.977	0.881	0.983	0.211	0.986	0.236	0.970	0.304	0.972	0.979	0.985	0.951	0.984	0.906	0.983
Humanities	0.567	0.355	0.624	0.307	0.580	0.271	-0.445	0.678	-0.087	0.921	-0.830	-0.436	0.935	0.885	0.966	0.947	0.929	0.966
Professional fields/other	0.794	0.865	0.774	0.859	0.771	0.861	0.330	0.958	0.442	0.897	0.039	0.960	0.829	0.872	0.650	0.799	0.465	0.702

APPENDIX B

Public Opinion on Evolution and Space

Data sources for the tables and figures in this appendix (each accessed 26 February 2018):

- Religious Landscape Study, http://www.pewforum.org/religious-landscape-study/.
- GSS Data Explorer: Interested in space exploration, https://gssdataexplorer.norc.org/variables/3459/vshow.
- GSS Data Explorer: Space exploration program, https://gssdataexplorer.norc.org/variables/199/vshow.
- GSS Data Explorer: Space exploration—version y,https://gssdataexplorer.norc.org/variables/199/vshow.
- Americans Continue to Rate NASA Positively, http://news.gallup.com/poll/102466/americans-continue-rate-nasa-positively.aspx.

Table B.1 VIEWS ON EVOLUTION BASED ON RELIGION; 2014 PEW RELIGIOUS LANDSCAPE STUDY

Religion	Evolved (natural selection)	Evolved (intelligent design)	Always existed in present form
Unaffiliated	63%	14%	15%
Catholic	31%	31%	29%
Evangelical Protestant	11%	25%	57%
Black Protestant	16%	31%	45%
Mainline Protestant	28%	31%	30%
Muslim	25%	25%	41%
Orthodox Christian	29%	25%	36%
Jewish	58%	18%	16%

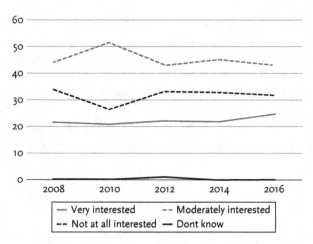

Figure B.1 Interest in space exploration issues from 2008 to 2014 based on the intspace GSS data set.
Credit Line: James S. J. Schwartz.

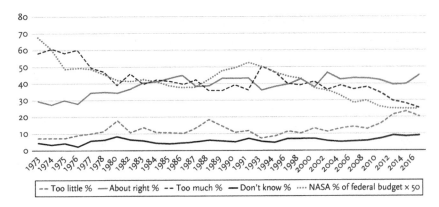

Figure B.2 Perceptions on space exploration spending from 1973 to 2016, based on combining totals from both the natspac (1973 onwards) and natspacy (1984 onwards) GSS data sets. NASA's share of the federal budget, which varied from 1.35% to 0.5% during this period, has been multiplied by 50 in order to better visualize a comparison of it with the results of the GSS questionnaire.

Credit Line: James S. J. Schwartz.

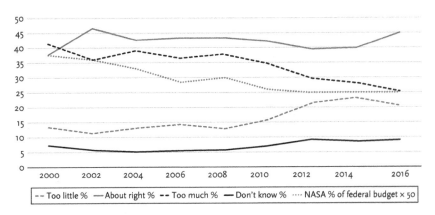

Figure B.3 Perceptions on space exploration spending from 2000 to 2016, based on both the natspac and natspacy GSS data sets. As in Fig. B.2, NASA's percentage of the federal budget, which varied from 0.75% to 0.5% during this period, has been scaled up by a factor of 50 in order to help visualize a comparison of it with the results of the GSS questionnaire.

Credit Line: James S. J. Schwartz.

Table B.2 A COMPARISON OF RESPONSES TO THE NATSPAC AND NATSPACY GSS QUESTIONNAIRES ON FUNDING FOR "SPACE EXPLORATION" (NATSPACY) VERSUS "THE SPACE EXPLORATION PROGRAM" (NATSPAC) DURING THE PERIOD FROM 1984 TO 2016 IN WHICH BOTH QUESTIONS WERE IN USE

	natspacy	natspac	Difference
Too little	13.80%	13.80%	0.01%
About right	41.20%	41.90%	0.66%
Too much	38.00%	37.30%	0.67%
Don't know	6.10%	6.70%	0.60%

REFERENCES

Ahlstrom-Vij, Kristoffer. 2013. In Defense of Veritistic Value Monism. *Pacific Philosophical Quarterly* 94: 19–40.

Almár, Iván. 2002. What Could COSPAR Do to Protect the Planetary and Space Environment? *Advances in Space Research* 30: 1577–1581.

Almár, Iván. 2010. New Concepts for an Advanced Planetary Protection Policy. In *Protecting the Environment of Celestial Bodies*, International Academy of Astronautics Cosmic Study, edited by Mahulena Hoffman, Petra Rettberg, and Mark Williamson, pp. 26–33. Available at: https://planetaryprotection.nasa.gov/file_download/57/PECBReport2010_IAAProtectingTheEnvmtOfCelestialBodies.pdf (accessed 10 July 2018).

Alston, William. 2006. *Beyond "Justification": Dimensions of Epistemic Justification*. New York: Cornell University Press.

Ambrosius, Joshua. 2015. Separation of Church and Space: Religious Influences on Public Support for U.S. Space Exploration Policy. *Space Policy* 32: 17–31.

Anderson, Elizabeth. 1993. *Value in Ethics and Economics*. Cambridge, MA: Harvard University Press.

Anthony, Niklas, and Reza Emami, M. 2018. Asteroid Engineering: The State-of-the-Art of Near-Earth Asteroids Science and Technology. *Progress in Aerospace Sciences* 100: 1–17.

Armstrong, Chris. 2013. Natural Resources: The Demands of Equality. *Journal of Social Philosophy* 44: 331–347.

Ashford, Elizabeth. 2007. The Duties Imposed by the Human Right to Basic Necessities. In *Freedom from Poverty as a Human Right: Who Owes What to the Very Poor?*, edited by Thomas Pogge, pp. 183–218. New York: Oxford University Press.

Baehr, Jason. 2012. Two Types of Wisdom. *Acta Analytica* 27: 81–97.

Bainbridge, William S. 2011. Cultural Beliefs About Extraterrestrials. In *Civilizations Beyond Earth: Extraterrestrial Life and Society*, edited by Douglas Vakoch and Albert Harrison, pp. 118–140. New York: Berghahn Books.

Bainbridge, William S. 2015. *The Meaning and Value of Spaceflight: Public Perceptions*. New York: Springer.

Balaguer, Mark. 2001. *Platonism and Anti-Platonism in Mathematics*. New York: Oxford University Press.

Banfield, Don, ed. 2018. Mars Scientific Goals, Objectives, Investigations, and Priorities: 2018 Version. Mars Exploration Program Analysis Group White Paper, available at: https://mepag.jpl.nasa.gov/reports.cfm (accessed 26 April 2019).

Baril, Anne. A Eudaimonist Approach to the Problem of Significance. *Acta Analytica* 25: 215–241.

Barker, Donald. 2015. The Mars Imperative: Species Survival and Inspiring a Globalized Culture. *Acta Astronautica* 107: 50–69.

Basu, Arindrajit, and Arthad Kurlekar. 2016. Highway to the Danger Zone: United States Legislative Framework Regulating the Commercial Space Sector. *Astropolitics* 14: 44–70.

Bayertz, Kurt. 2006. Three Arguments for Scientific Freedom. *Ethical Theory and Moral Practice* 9: 377–398.

Bergström, Lars. 1987. On the Value of Scientific Knowledge. *Grazer Philosophische Studien* 30: 53–63.

Bergström, Lars. 1995. Notes on the Value of Science. In *Logic, Methodology and Philosophy of Science IX*, edited by Dag Prawitz, Brian Skyrms, and Dag Westerstål, pp. 499–522. New York: Elsevier Science.

Billings, Linda. 2006. How Shall We Live in Space? Culture, Law and Ethics in Spacefaring Society. *Space Policy* 22: 249–255.

Bjørnvig, Thore. 2013. Outer Space Religion and the Overview Effect: A Critical Inquiry into a Classic of the Pro-Space Movement. *Astropolitics* 11: 4–24.

Bondy, Patrick. 2018. Epistemic Value. *Internet Encyclopedia of Philosophy.* Available at: https://www.iep.utm.edu/ep-value/ (accessed 25 August 2018).

Bradley, Ben. 2006. Two Concepts of Intrinsic Value. *Ethical Theory and Moral Practice* 9: 111–130.

Brearly, Andrew. 2006. Mining the Moon: Owning the Night Sky? *Astropolitics* 4: 43–67.

Brown, Mark, and David Guston. 2009. Science, Democracy, and the Right to Research. *Science and Engineering Ethics* 15: 351–366.

Brown, Matthew. 2010. Genuine Problems and the Significance of Science. *Contemporary Pragmatism* 7: 131–153.

Bush, Vannevar. 1945. *Science, the Endless Frontier.* Washington, DC: United States Government Printing Office.

Callicott, J. Baird. 1986. Moral Considerability and Extraterrestrial Life. In *Beyond Spaceship Earth: Environmental Ethics and the Solar System*, edited by Eugene Hargrove, pp. 227–259. San Francisco: Sierra Club Books.

Carter, J. Adam. 2011. Kvanvig on Pointless Truths and the Cognitive Ideal. *Acta Analytica* 26: 285–293.

Carter, J. Adam, and Emma Gordon. 2014. Objectual Understanding and the Value Problem. *American Philosophical Quarterly* 51: 1–13.

Chen, Chuansheng, et al. 1999. Population Migration and the Variation of Dopamine D4 Receptor (DRD4) Allele Frequencies Around the Globe. *Evolution & Human Behavior* 20: 309–324.

Ciani, Andrea, Shany Edelman, and Richard Ebstein. 2013. The Dopamine D4 Receptor (DRD4) Exon 3 VNTR Contributes to Adaptive Personality Differences in an Italian Small Island Population. *European Journal of Personality* 27: 593–604.

Cicovacki, Predrag. 2017. On the Puzzling Value of Human Life. *Ethics & Bioethics (in Central Europe)* 7: 155–168.

Cockell, Charles. 2004. The Value of Humans in the Biological Exploration of Space. *Earth, Moon, and Planets* 94: 233–243.

Cockell, Charles. 2005a. The Value of Microorganisms. *Environmental Ethics* 27: 375–390.

Cockell, Charles. 2005b. Planetary Protection—A Microbial Ethics Approach. *Space Policy* 21: 287–292.

Cockell, Charles. 2006. The Ethical Relevance of Earth-like Extrasolar Planets. *Environmental Ethics* 28: 303–314.

Cockell, Charles. 2007. *Space On Earth: Saving Our World by Seeking Others.* New York: Palgrave Macmillan.

Cockell, Charles. 2008a. Interstellar Planetary Protection. *Advances in Space Research* 42: 1161–1165.

Cockell, Charles. 2008b. An Essay on Extraterrestrial Liberty. *Journal of the British Interplanetary Society* 61: 255–275.

Cockell, Charles. 2010. Astrobiology—What Can We Do on the Moon? *Earth, Moon, and Planets* 107: 3–10.

Cockell, Charles. 2016a. The Ethical Status of Microbial Life on Earth and Elsewhere: In Defense of Intrinsic Value. In *The Ethics of Space Exploration*, edited by James S. J. Schwartz and Tony Milligan, pp. 167–179. New York: Springer.

Cockell, Charles. 2016b. Disobedience in Outer Space. In *Dissent, Revolution, and Liberty Beyond Earth*, edited by Charles Cockell, pp. 21–40. New York: Springer.

Cockell, Charles, and Horneck, Gerda. 2004. A Planetary Park System for Mars. *Space Policy* 20: 291–295.

Cockell, Charles, and Horneck, Gerda. 2006. Planetary Parks—Formulating a Wilderness Policy for Planetary Bodies. *Space Policy* 22: 256–261.

Cook, Summer, Marvin Druger, and Lori Ploutz-Snyder. 2011. Scientific Literacy and Attitudes toward American Space Exploration Among College Undergraduates. *Space Policy* 27: 48–52.

Cooper, Lawrence. 2003. Encouraging Space Exploration Through a New Application of Space Property Rights. *Space Policy* 19: 111–118.

Cooper, P. David. 2016. Babies on Mars: Biomedical Considerations for the First Martian Generation. *Journal of the British Interplanetary Society* 69: 307–325.

Crawford, Ian. 2001. The Scientific Case for Human Space Exploration. *Space Policy* 17: 155–159.

Crawford, Ian. 2005. Towards an Integrated Scientific and Social Case for Human Space Exploration. *Earth, Moon, and Planets* 94: 245–266.

Crawford, Ian. 2012. Dispelling the Myth of Robotic Efficiency. *Astronomy and Geophysics* 53: 2.22–2.26.

Crawford, Ian. 2015. Lunar Resources: A Review. *Progress in Physical Geography* 39: 137–167.

Crawford, Ian. 2016. The Long-term Scientific Benefits of a Space Economy. *Space Policy* 37: 58–61.

Crawford, Ian, et al. 2012. Back to the Moon: The Scientific Rationale for Resuming Lunar Surface Exploration. *Planetary and Space Science* 74: 3–14.

Dahl, Robert. 1985. *A Preface to Economic Democracy.* Berkeley, CA: University of California Press.

Dark, Taylor. 2007. Reclaiming the Future: Space Advocacy and the Idea of Progress. In *Societal Impact of Spaceflight*, edited by Steven Dick and Roger Launius, pp. 555–571. NASA SP-2007-4801.

De Man, Philip. 2019. Interpreting the UN Space Treaties as the Basis for a Sustainable Regime of Space Resource Exploitation. In *The Space Treaties at Crossroads*, edited by G. D. Kryiakopolous and M. Manoli, pp. 15–33. New York: Springer.

de Regt, Henk. 2017. *Understanding Scientific Understanding.* New York: Oxford University Press.

Dech, Stefan. 2006. The Earth Surface. In *Utilization of Space: Today and Tomorrow*, edited by B. Feuerbacher and H. Stoewer, pp. 53–90. New York: Springer.

Dellsén, Finnur. 2016. Scientific Progress: Knowledge Versus Understanding. *Studies in History and Philosophy of Science* 56: 72–83.

Dettmar, Ralf-Jürgen. 2006. Astronomy and Astrophysics. In *Utilization of Space: Today and Tomorrow*, edited by B. Feuerbacher and H. Stoewer, pp. 169–194. New York: Springer.

Dittus, Hansjörg. 2006. Fundamental Physics. In *Utilization of Space: Today and Tomorrow*, edited by B. Feuerbacher and H. Stoewer, pp. 275–296. New York: Springer.

Douglas, Heather. 2014. Pure Science and the Problem of Progress. *Studies in History and Philosophy of Science* 46: 55–63.

Dundas, Colin, et al. 2018. Exposed Subsurface Ice Sheets in the Martian Mid-Latitudes. *Science* 359: 199–201.

von der Dunk, Frans. 2018. Private Property Rights and the Public Interest in Exploration of Outer Space. *Biological Theory* 13: 142–151.

Ehrenfreund, Pascale, Margaret Race, and David Labdon. 2013. Responsible Space Exploration and Use: Balancing Stakeholder Interests. *New Space* 1: 60–72.

Elgin, Catherine. 2017. *True Enough*. Cambridge, MA: MIT Press.

ElGindi, Tamer. 2017. Natural Resource Dependency, Neoliberal Globalization, and Income Inequality: Are They Related? A Longitudinal Study of Developing Countries (1980–2010). *Current Sociology* 65: 21–53.

Elvis, Martin. 2014. How Many Ore-Bearing Asteroids? *Planetary and Space Science* 91: 20–26.

Elvis, Martin. 2016. What can Space Resources do for Astronomy and Planetary Science? *Space Policy* 37: 65–76.

Elvis, Martin, Tony Milligan, and Alanna Krolikowski. 2016. The Peaks of Eternal Light: A Near-term Property Issue on the Moon. *Space Policy* 38: 30–38.

Foster, Jamie, and Jennifer Drew. 2009. Astrobiology Undergraduate Education: Students' Knowledge and Perceptions of the Field. *Astrobiology* 9: 325–333.

Frankel, Mark. 2009. Private Interests Count Too, commentary on "Science, Democracy, and the Right to Research," by Mark Brown and David Guston. *Science and Engineering Ethics* 15: 367–373.

Freitas, Robert. 1980. A Self-Reproducing Interstellar Probe. *Journal of the British Interplanetary Society* 33: 251–264.

Gerzer, Rupert, Ruth Hemmersbach, and Gerda Horneck. 2006. Life Science. In *Utilization of Space: Today and Tomorrow*, edited by B. Feuerbacher and H. Stoewer, pp. 341–373. New York: Springer.

Gettier, Edmund. 1963. Is Justified True Belief Knowledge? *Analysis* 23: 121–123.

Gilster, Paul. 2013. The Interstellar Vision: Principles and Practice. *Journal of the British Interplanetary Society* 66: 223–232.

Goldfarb, Jillian, and Douglas Kriner. 2017. Building Public Support for Science Spending: Misinformation, Motivated Reasoning, and the Power of Corrections. *Science Communication* 39: 77–100.

Goldman, Alvin. 1999. *Knowledge in a Social World*. Oxford: Clarendon Press.

Gören, Erkan. 2016. The Biogeographic Origins of Novelty-Seeking Traits. *Evolution & Human Behavior* 37: 456–469.

Gottlieb, Joseph. 2019. Space Colonization and Existential Risk. *Journal of the American Philosophical Association* 5: 306–320.

Grimm, Stephen. 2011a. What Is Interesting? *Logos and Episteme* 2: 515–542.

Grimm, Stephen. 2011b. Understanding. In *The Routledge Companion to Epistemology*, edited by Sven Bernecker and Duncan Pritchard, pp. 84–94. New York: Routledge.

Grimm, Stephen. 2012. The Value of Understanding. *Philosophy Compass* 7: 103–117.

Gros, Claudius. 2016. Developing Ecospheres on Transiently Habitable Planets: The Genesis Project. *Astrophysics and Space Science* 361: 324.

Gros, Claudius. 2019. Why Planetary and Exoplanetary Protection Differ: The Case of Long Duration Genesis Missions to Habitable but Sterile M-Dwarf Oxygen Planets. *Acta Astronautica* 157: 263–267.

Gurtuna, Ozgur. 2013. *Fundamentals of Space Business and Economics*. New York: Springer Briefs in Space Development.

Hansson, Lena, and Andreas Redfors. 2013. Lower Secondary Students' Views in Astrobiology. *Research in Science Education* 43: 1957–1978.

Harris, Alan, and Germano D'Abramo. 2015. The Population of Near-Earth Asteroids. *Icarus* 257: 302–312.

Hartmann, William. 1984. Space Exploration and Environmental Issues. *Environmental Ethics* 6: 227–239.

Hein, Andreas, et al. 2012. World Ships—Architectures and Feasibility Revisited. *Journal of the British Interplanetary Society* 65: 119–133.

Hickman, John. 2010. Extraterrestrial National Territory and the International System. *Astropolitics* 8: 62–71.

Horwich, Paul. 2006. The Value of Truth. *Noûs* 40: 347–360.

Jakhu, Ram. 2017. Regulatory Process for Communications Satellite Frequency Allocations. In *Handbook of Satellite Applications*, edited by Joseph Pelton, Scott Madry, and Sergio Camacho-Lara, pp. 359–381. New York: Springer.

Jakhu, Ram, and Maria Buzdugan. 2008. Development of the Natural Resources of the Moon and Other Celestial Bodies: Economic and Legal Aspects. *Astropolitics* 6: 201–250.

Jakosky, Bruce. 2006. *Science, Society, and the Search for Life in the Universe*. Tucson, AZ: University of Arizona Press.

Janvid, Mikael. 2012. Knowledge Versus Understanding: The Cost of Avoiding Gettier. *Acta Analytica* 27: 183–197.

Jedicke, Robert, et al. 2018. Availability and Delta-v Requirements for Delivering Water Extracted From Near-Earth Objects to Cis-Lunar Space. *Planetary and Space Science* 159: 28–42.

Kant, Immanuel. 1993. *Grounding for the Metaphysics of Morals*. 3rd ed. Translated by James W. Ellington. Indianapolis and Cambridge: Hackett Publishing Company.

Kavka, Gregory. 1982. The Paradox of Future Individuals. *Philosophy & Public Affairs* 11: 93–112.

Kennedy, Robert, Kenneth Roy, and David Fields. 2013. Dyson Dots: Changing the Solar Constant to a Variable With Photovoltaic Lightsails. *Acta Astronautica* 82: 225–237.

Kidd, Celeste, and Benjamin Hayden. 2015. The Psychology and Neuroscience of Curiosity. *Neuron* 88: 449–460.

Kitcher, Philip. 2001. *Science, Truth, and Democracy*. Rev. ed. New York: Oxford University Press.

Kitcher, Philip. 2004. On the Autonomy of the Sciences. *Philosophy Today* 48: 51–57.

Kitcher, Philip. 2011. *Science in a Democratic Society*. Amherst, NY: Prometheus Books.

Kluger, A. N., Z. Siegfried, and Richard Ebstein. 2002. A Meta-analysis of the Association Between DRD4 Polymorphism and Novelty Seeking. *Molecular Psychology* 7: 712–717.

Kminek, G., et al. 2017. COSPAR's Planetary Protection Policy. *Space Research Today* 200: 12–25.

Koepsell, David. 2017. Mars One: Human Subject Concerns? *Astropolitics* 15: 97–111.

Kondo, Yoji, et al., eds. 2003. *Interstellar Travel and Multi-Generation Space Ships.* Burlington, Ontario: Apogee Books.

Kramer, William. 2014. Extraterrestrial Environmental Impact Assessments—A Foreseeable Prerequisite for Wise Decisions Regarding Outer Space Exploration, Research and Development. *Space Policy* 30: 215–222.

Kraut, Richard. 2018. Aristotle's Ethics. *The Stanford Encyclopedia of Philosophy.* Summer 2018 ed. Edited by Edward Zalta. Available at: https://plato.stanford. edu/archives/sum2018/entries/aristotle-ethics/ (accessed 26 September 2019).

Krolikowski, Alanna, and Martin Elvis. 2019. Making Policy for New Asteroid Activities: In Pursuit of Science, Settlement, Security, or Sales? *Space Policy* 47: 7–17.

Kvanvig, Jonathan. 2003. *The Value of Knowledge and the Pursuit of Understanding.* New York: Cambridge University Press.

Kvanvig, Jonathan. 2008. Pointless Truth. *Midwest Studies in Philosophy* 32: 199–212.

Kvanvig, Jonathan. 2013. Curiosity and the Response-Dependent Special Value of Understanding. In *Knowledge, Virtue, and Action: Essays on Putting Epistemic Virtues to Work*, edited by Tim Hening and David Schweikard, pp. 151–174. New York: Routledge.

Lahti, Jari, et al. 2006. Socio-demographic Characteristics Moderate the Association Between DRD4 and Novelty Seeking. *Personality and Individual Differences* 40: 533–543.

Langston, Sara. 2016. Space Travel: Risk, Ethics, and Governance in Commercial Human Spaceflight. *New Space* 4: 83–97.

Laucht, Manfred, Katja Becker, and Martin Schmidt. 2006. Visual Exploratory Behavior in Infancy and Novelty Seeking in Adolescence: Two Developmentally Specific Phenotypes of DRD4? *Journal of Child Psychology and Psychiatry* 47: 1143–1151.

Launius, Roger. 2003. Public Opinion Polls and Perceptions of US Human Spaceflight. *Space Policy* 19: 163–175.

Lee, Keekok. 1994. Awe and Humility: Intrinsic Value in Nature—Beyond an Earthbound Environmental Ethics. In *Philosophy and the Natural Environment*, edited by Robin Attfield and Andrew Belsey, 89–101. New York: Cambridge University Press.

Lee, Ricky. 2012. *Law and Regulation of Commercial Mining of Minerals in Outer Space.* New York: Springer.

Leopold, Aldo. 1949. *A Sand County Almanac, and Sketches Here and There.* New York: Oxford University Press.

Lester, Daniel, and Michael Robinson. 2009. Visions of Exploration. *Space Policy* 25: 236–243.

Lester, Daniel, and Harley Thronson. 2011. Human Space Exploration and Human Spaceflight: Latency and the Cognitive Scale of the Universe. *Space Policy* 27: 89–93.

Lewis, John. 1997. *Mining the Sky: Untold Riches from the Asteroids, Comets, and Planets.* New York:Basic Books.

Lewis, John. 2015. *Asteroid Mining 101: Wealth for the New Space Economy.* San Jose, CA:Deep Space Industries.

Liberto, Hallie. 2014. The Exploitation Solution to the Non-Identity Problem. *Philosophical Studies* 167: 73–88.

Limerick, Patricia. 1992. Imagined Frontiers: Westward Expansion and the Future of the Space Program. In *Space Policy Alternatives*, edited by Radford Byerly, pp. 249–262. Boulder, CO: Westview Press.

Loewenstein, George. 1994. The Psychology of Curiosity: A Review and Reinterpretation. *Psychological Bulletin* 116: 75–98.

Lubin, Philip. 2016. A Roadmap to Interstellar Flight. *Journal of the British Interplanetary Society* 69: 40–72.

Lynch, Michael. 2004. *True to Life: Why Truth Matters*. Cambridge, MA: MIT Press.

Lynch, Michael. 2017. Understanding and Coming to Understand. In *Making Sense of the World: New Essays on the Philosophy of Understanding*, edited by Stephen Grimm, pp. 194–208. New York: Oxford University Press.

Maddy, Penelope. 2005. Three Forms of Naturalism. In *The Oxford Handbook of Philosophy of Mathematics and Logic*, edited by Stewart Shapiro, pp. 437–459. New York: Oxford University Press.

Maddy, Penelope. 2007. *Second Philosophy: A Naturalistic Method*. New York: Oxford University Press.

Maddy, Penelope. 2011. *Defending the Axioms: On the Philosophical Foundations of Set Theory*. New York: Oxford University Press.

Mankins, John, Willa Mankins, and Helen Walter. 2018. Biological Challenges of True Space Settlement. *Acta Astronautica* 146: 378–386.

Marin, Frédéric, and Camille Beluffi. 2018. Computing the Minimal Crew for a Multi-Generational Space Journey Towards Proxima Centauri b. *Journal of the British Interplanetary Society* 71: 45–52.

Marshall, Alan. 1993. Ethics and the Extraterrestrial Environment. *Journal of Applied Philosophy* 10: 227–236.

Marshall, Alan. 1995. Development and Imperialism in Space. *Space Policy* 11: 41–52.

Matloff, Greg, C. Bangs, and Les Johnson. 2014. *Harvesting Space for a Greener Earth*. 2nd ed. New York: Springer.

Matthews, Jack, and Sean McMahon. 2018. Exogeoconservation: Protecting Geological Heritage on Celestial Bodies. *Acta Astronautica* 149: 55–60.

Matthews, Luke, and Paul Butler. 2011. Novelty-Seeking DRD4 Polymorphisms are Associated with Human Migration Distance Out-of-Africa After Controlling for Neutral Population Gene Structure. *American Journal of Physical Anthropology* 145: 382–389.

McCurdy, Howard. 2011. *Space and the American Imagination*. 2nd ed. Baltimore: Johns Hopkins University Press.

McInnes, Colin. 2016. Near Earth Asteroid Resource Utilisation for Large In-Orbit Reflectors. *Space Policy* 37: 62–64.

McKay, Christopher. 1990. Does Mars Have Rights? An Approach to the Environmental Ethics of Planetary Engineering. In *Moral Expertise: Studies in Practical and Professional Ethics*, edited by Don MacNiven, pp. 184–197. London: Routledge.

McMahon, Sean. 2016. The Aesthetic Objection to Terraforming Mars. In *The Ethics of Space Exploration*, edited by James S. J. Schwartz and Tony Milligan, pp. 209–218. New York: Springer.

Melanson, William. 2012. You Can't Buy Much with Intellectual Credit. *American Philosophical Quarterly* 49: 253–265.

Metzger, Philip. 2016. Space Development and Space Science Together, an Historic Opportunity. *Space Policy* 37: 77–91.

Meyer, Lukas. 2016. Intergenerational Justice. *Stanford Encyclopedia of Philosophy*. Summer 2016 ed. Edited by Edward Zalta. Available at: https://plato.stanford.edu/archives/sum2016/entries/justice-intergenerational/ (accessed 22 September 2018).

Milligan, Tony. 2011. Property Rights and the Duty to Extend Human Life. *Space Policy* 27: 190–193.

Milligan, Tony. 2013. Scratching the Surface: The Ethics of Helium-3 Extraction. Paper delivered to the 8th International Academy of Astronautics Symposium on the Future of Space Exploration: Towards the Stars, July 3–5, 2013, Torino, Italy.

Milligan, Tony. 2015a. *Nobody Owns the Moon: The Ethics of Space Exploitation*. Jefferson, NC: McFarland and Company.

Milligan, Tony. 2015b. Asteroid Mining, Integrity and Containment. In *Commercial Space Exploration: Ethics, Policy and Governance*, edited by Jai Galliott, pp. 123–134. Dorchester, UK: Ashgate.

Milligan, Tony. 2015c. Rawlsian Deliberation About Space Settlement. In *Human Governance Beyond Earth*, edited by Charles Cockell, pp. 9–22. New York: Springer.

Milligan, Tony. 2018. Valuing Humans and Valuing Places: "Integrity" and the Preferred Terminology for Geoethics. *Geosciences* 8: 1–14.

Moore, G. E. 1903. *Principia Ethica*. New York: Cambridge University Press.

Moore, John. 2003. Kin-Based Crews for Interstellar Multi-Generation Space Travel. In *Interstellar Travel and Multi-Generation Space Ships*, edited by Yoji Kondo et al., pp. 80–88. Burlington, Ontario: Apogee Books.

Mousis, Olivier, et al. 2014. Scientific Rationale for Saturn's *in situ* Exploration. *Planetary and Space Science* 104: 29–47.

Munafò, Marcus, et al. 2008. Association of the Dopamine D4 Receptor (DRD4) Gene and Approach-Related Personality Traits: Meta-Analysis and New Data. *Biological Psychiatry* 63: 197–206.

Munévar, Gonzalo. 1998. *Evolution and the Naked Truth: A Darwinian Approach to Philosophy*. Burlington, VA: Ashgate.

Munévar, Gonzalo. 2014. Space Exploration and Human Survival. *Space Policy* 30: 197–201.

Munévar, Gonzalo. 2016. Space Colonies and Their Critics. In *The Ethics of Space Exploration*, edited by James S. J. Schwartz and Tony Milligan, pp. 31–45. New York: Springer.

Murchie, Scott, Daniel Britt, and Carle Pieters. 2014. The Value of Phobos Sample Return. *Planetary and Space Science* 102: 176–182.

Musk, Elon. 2016. Making Humans a Multi-Planetary Species. *New Space* 5: 46–61.

Nadeau, François. 2013. Explaining Public Support for Space Exploration Funding in America: A Multivariate Analysis. *Acta Astronautica* 86: 158–166.

National Geographic Society. 1994. Proceedings of the *What Is the Value of Space Exploration?* Symposium. Available at: https://www.hq.nasa.gov/office/hqlibrary/documents/o33273018.pdf (accessed 7 July 2016).

Neta, Ram. 2008. How to Naturalize Epistemology. In *New Waves in Epistemology*, edited by Vincent Hendricks and Duncan Pritchard, pp. 324–353. New York: Palgrave MacMillan.

Newell, Catherine. 2014. Without Having Seen: Faith, the Future, and the Final American Frontier. *Astropolitics* 12: 148–166.

Newman, Christopher, and Mark Williamson. 2018. Space Sustainability: Reframing the Debate. *Space Policy* 46: 30–37.

Nicogossian, Arnauld, et al., eds. 2016. *Space Physiology and Medicine*. New York: Springer.

Norton, Bryan. 1987. *Why Preserve Natural Variety?* Princeton, NJ: Princeton University Press.

Nozick, Robert. 1981. *Philosophical Explanations*. Cambridge, MA: Harvard University Press.

O'Neill, John. 1992. The Varieties of Intrinsic Value. *The Monist* 75: 119–137.

Oreiro, Raquel, and Jordi Solbes. 2017. Secondary School Students' Knowledge and Opinions on Astrobiology Topics and Related Social Issues. *Astrobiology* 17: 91–99.

Parfit, Derek. 1982. Future Generations: Further Problems. *Philosophy & Public Affairs* 11: 113–172.

Parfit, Derek. 1984. *Reasons and Persons*. Oxford: Clarendon Press.

Perry, Ian. 2017. Law of Space Resources and Operations on Celestial Bodies: Implications for Legislation in the United States. *Astropolitics* 15: 1–26.

Persson, Erik. 2012. The Moral Status of Extraterrestrial Life. *Astrobiology* 12: 976–984.

Persson, Erik, Klara Capova, and Yuan Li. 2019. Attitudes Towards the Scientific Search for Extraterrestrial Life Among Swedish High School and University Students. *International Journal of Astrobiology* 18: 280–288.

Peters, Ted. 2013. Would the Discovery of ETI Provoke a Religious Crisis? In *Astrobiology, History, and Society*, edited by Douglas Vakoch, pp. 341–355. New York: Springer.

Pettinico, George. 2011. American Attitudes About Life Beyond Earth: Beliefs, Concerns, and the Role of Education and Religion in Shaping Public Perceptions. In *Civilizations Beyond Earth: Extraterrestrial Life and Society*, edited by Douglas Vakoch and Albert Harrison, 102–117. New York: Berghahn Books.

Pilchman, Daniel. 2015. Three Ethical Perspectives on Asteroid Mining. In *Commercial Space Exploration: Ethics, Policy and Governance*, edited by Jai Galliott, pp. 134–147. Dorchester, UK: Ashgate.

Plantinga, Alvin. 1993. *Warrant and Proper Function*. New York: Oxford University Press.

Pop, Virgiliu. 2012. The Property Status of Lunar Resources. In *Moon: Prospective Energy and Material Resources*, edited by Viorel Badescu, pp. 553–564. New York: Springer.

Post, Robert. 2009. Constitutional Restraints on the Regulations of Scientific Speech and Scientific Research, commentary on "Democracy, Individual Rights and the Regulation of Science," by J. Weinstein. *Science and Engineering Ethics* 15: 431–438.

Pritchard, Duncan. 2010. Knowledge and Understanding. In *The Nature and Value of Knowledge: Three Investigations*, by Duncan Pritchard, Alan Millar, and Adrian Haddock, pp. 1–88. New York: Oxford University Press.

Pritchard, Duncan. 2016. Epistemic Axiology. In *Epistemic Reasons, Norms and Goals*, edited by Martin Grajner and Pedro Schmechtig, pp. 407–422. Berlin: De Gruyter.

Pyne, Stephen. 2006. Seeking Newer Worlds: An Historical Context for Space Exploration. In *Critical Issues in the History of Spaceflight*, edited by Steven Dick and Roger Launius, pp. 7–35. NASA SP-2006-4702.

Randolph, Richard, and Christopher McKay. 2014. Protecting and Expanding the Richness and Diversity of Life, an Ethic for Astrobiology Research and Space Exploration. *International Journal of Astrobiology* 13: 28–34.

Ratke, Lorenz. 2006. Materials Sciences. In *Utilization of Space: Today and Tomorrow*, edited by B. Feuerbacher and H. Stoewer, pp. 297–340. New York: Springer.

Rawls, John. 1971. *A Theory of Justice*. Cambridge, MA: Belknap Press.

Regis, Edward. 1985. The Moral Status of Multigenerational Interstellar Exploration. In *Interstellar Migration and the Human Experience*, edited by Ben Finney and Eric Jones, pp. 248–259. Berkeley, CA: University of California Press.

Reiman, Saara. 2009. Is Space an Environment? *Space Policy* 25: 81–87.

Reiman, Saara. 2010. On Sustainable Exploration of Space and Extraterrestrial Life. *Journal of Cosmology* 12: 3894–3903.

Reiman, Saara. 2011. Sustainability in Space Exploration—An Ethical Perspective. Lakewood, CO: Mars Society Papers.

Reio, Thomas. 2012. Curiosity and Exploration. In *Encyclopedia of the Sciences of Learning*, edited by Norbert Seel, pp. 894–896. New York: Springer.

Rendall, Matthew. 2011. Non-Identity, Sufficiency and Exploitation. *The Journal of Political Philosophy* 19: 229–247.

Rivkin, Andrew, and Francesca DeMeo. 2019. How Many Hydrated NEOs Are There? *Journal of Geophysical Research: Planets* 124: 128–142.

Roberts, M. A. 2015. The Nonidentity Problem. *The Stanford Encyclopedia of Philosophy*. Winter 2015 ed. Edited by Edward Zalta. Available at: https://plato.stanford.edu/archives/win2015/entries/nonidentity-problem/ (accessed 22 September 2018).

Rolston, Holmes, III. 1986. The Preservation of Natural Value in the Solar System. In *Beyond Spaceship Earth: Environmental Ethics and the Solar System*, edited by Eugene Hargrove, pp. 140–182. San Fransisco: Sierra Club Books.

Roussos, Panos, Stella Giakoumaki, and Panos Bitsios. 2009. Cognitive and Emotional Processing in High Novelty Seeking Associated with the L-DRD4 Genotype. *Neuropsychologia* 47: 1654–1659.

Rummel, John. 2019. From Planetary Quarantine to Planetary Protection: A NASA and International Story. *Astrobiology* 19: 624–627.

Rummel, John, Margaret Race, and Gerda Horneck. 2012. Ethical Considerations for Planetary Protection: A Workshop. *Astrobiology* 12: 1017–1023.

Rummel, John, et al. 2010. The Integration of Planetary Protection Requirements and Medical Support on a Mission to Mars. In *Colonizing Mars: The Human Mission to the Red Planet*, edited by Joel Levine and Rudolf Schild, pp. 164–171. Cambridge, MA: Cosmology Science Publishers.

Ryan, Sharon. 2012. Widsom, Knowledge and Rationality. *Acta Analytica* 27: 99–112.

Sadeh, Eligar. 2015. Impacts of the Apollo Program on NASA, the Space Community, and Society. In *Historical Studies of the Societal Impact of Spaceflight*, edited by Steven Dick, 491–534. NASA SP-2015-4803.

Sagan, Carl. 1994. *Pale Blue Dot: A Vision of the Human Future in Space*. New York: Random House.

Sagan, Carl. 2013. *Cosmos*. Rev. ed. New York: Ballantine Books.

Saletta, Morgan, and Kevin Orrman-Rossiter. 2018. Can Space Mining Benefit All of Humanity?: The Resource Fund and Citizen's Dividend Model of Alaska, the "Last Frontier." *Space Policy* 43: 1–6.

Sanchez, Joan-Pau, and Colin McInnes. 2011. Asteroid Resource Map for Near-Earth Space. *Journal of Spacecraft and Rockets* 48: 153–165.

Sanchez, Joan-Pau, and Colin McInnes. 2013. Available Asteroid Resources in the Earth's Neighbourhood. In *Asteroids: Prospective Energy and Material Resources*, edited by Viorel Badescu, pp. 439–458. New York: Springer.

Sarewitz, Daniel, et al. 2004. Science Policy in Its Social Context. *Philosophy Today* 48: 67–83.

Savulescu, Julian, and Richard Momeyer. 1997. Should Informed Consent Be Based on Rational Beliefs? *Journal of Medical Ethics* 23: 282–288.

Scanlon, T. M. 1998. *What We Owe to Each Other*. Cambridge, MA: Harvard University Press.

Schinka, J. A., E. A. Letsch, and F. C. Crawford. 2003. DRD4 and Novelty Seeking: Results of Meta-Analyses. *American Journal of Medical Genetics* 114: 643–648.

Schrogl, K.U., et al., eds. 2015. *Handbook of Space Security.* New York: Springer.

Schulze-Makuch, Dirk, and Paul Davies. 2013. Destination Mars: Colonization via Initial One-Way Missions. *Journal of the British Interplanetary Society* 66: 11–14.

Schwartz, James. 2011. Our Moral Obligation to Support Space Exploration. *Environmental Ethics* 33: 67–88.

Schwartz, James. 2013a. *Nominalism in Mathematics: Modality and Naturalism.* Doctoral Dissertation, Wayne State University.

Schwartz, James. 2013b. On the Moral Permissibility of Terraforming. *Ethics and the Environment* 18: 1–31.

Schwartz, James. 2014. Prioritizing Scientific Exploration: A Comparison of the Ethical Justifications for Space Development and for Space Science. *Space Policy* 30: 202–208.

Schwartz, James. 2015a. Mathematical Structuralism, Modal Nominalism, and the Coherence Principle. *Philosophia Mathematica* 23: 367–385.

Schwartz, James. 2015b. Rendezvous With Research: Government Support of Science in a Space Society. In *Human Governance Beyond Earth*, edited by Charles Cockell, pp. 197–211. New York: Springer.

Schwartz, James. 2015c. Fairness as a Moral Grounding for Space Policy. In *The Meaning of Liberty Beyond Earth*, edited by Charles Cockell, pp. 69–89. New York: Springer.

Schwartz, James. 2016a. On the Methodology of Space Ethics. In *The Ethics of Space Exploration*, edited by James S. J. Schwartz and Tony Milligan, pp. 93–107. New York: Springer.

Schwartz, James. 2016b. Near-Earth Water Sources: Ethics and Fairness. *Advances in Space Research* 58: 402–407.

Schwartz, James. 2017a. Myth-Free Space Advocacy Part I: The Myth of Innate Exploratory and Migratory Urges. *Acta Astronautica* 137: 450–460.

Schwartz, James. 2017b. Myth-Free Space Advocacy Part II: The Myth of the Space Frontier. *Astropolitics* 15: 167–184.

Schwartz, James. 2018a. Myth-Free Space Advocacy Part III: The Myth of Educational Inspiration. *Space Policy* 43: 24–32.

Schwartz, James. 2018b. Worldship Ethics: Obligations to the Crew. *Journal of the British Interplanetary Society* 71: 53–64.

Schwartz, James. 2019a. Space Settlement: What's the Rush? *Futures* 110: 56–59.

Schwartz, James. 2019b. Where No Planetary Protection Policy Has Gone Before. *International Journal of Astrobiology* 18: 353–361.

Schwartz, James. 2019c. Mars: Science Before Settlement. *Theology and Science* 17: 324–331.

Schwartz, James. forthcoming. Myth-Free Space Advocacy Part IV: The Myth of Public Support for Astrobiology. In *Social and Conceptual Issues in Astrobiology*, edited by Kelly Smith and Carlos Mariscal. New York: Oxford University Press.

Schwartz, James, and Tony Milligan. 2017. Some Ethical Constraints on Near-Earth Resource Exploitation. In *Yearbook on Space Policy 2015*, edited by Cenan Al-Ekabi et al., pp. 227–239. New York: Springer.

Shelhamer, Mark. 2017. Why Send Humans Into Space? Science and Non-Science Motivations for Human Space Flight. *Space Policy* 42: 37–40.

Shepard, Michael. 2015. *Asteroids: Relics of Ancient Time.* Cambridge, UK: Cambridge University Press.

Shue, Henry. 1996. *Basic Rights: Subsistence, Affluence, and U.S. Foreign Policy*. 2nd ed. Princeton, NJ: Princeton University Press.

Simon, Jeremy. 2006. The Proper Ends of Science: Philip Kitcher, Science, and the Good. *Philosophy of Science* 73: 194–214.

Slobodian, Rayna. 2015. Selling Space Colonization and Immortality: A Psychosocial, Anthropological Critique of the Rush to Colonize Mars. *Acta Astronautica* 113: 89–104.

Smith, Kelly. 2009. The Trouble With Intrinsic Value: An Ethical Primer. In *Exploring the Origin, Extent, and Future of Life*, edited by Constance Bertka, pp. 261–280. New York: Cambridge University Press.

Smith, Kelly. 2016a. The Curious Case of the Martian Microbes: Mariomania, Intrinsic Value and the Prime Directive. In *The Ethics of Space Exploration*, edited by James S. J. Schwartz and Tony Milligan, pp. 195–208. New York: Springer.

Smith, Kelly. 2016b. Cultural Evolution and the Colonial Imperative. In *Dissent, Revolution and Liberty Beyond Earth*, edited by Charles Cockell, pp. 169–187. New York: Springer.

Snow, Catherine, and Kenne Dibner, eds. 2016. *Scientific Literacy: Concepts, Contexts, and Consequences*. Washington, DC: National Academies Press.

Solomon, Lewis. 2008. *The Privatization of Space Exploration: Business, Technology, Law and Policy*. Piscataway, NJ: Transaction Publishers.

Sorenson, Roy. 2011. Interestingly Dull Numbers. *Philosophy and Phenomenological Research* 82: 655–673.

Sosa, Ernest. 2001. For the Love of Truth? In *Virtue Epistemology: Essays on Epistemic Virtue and Responsibility*, edited by Abrol Fairweather and Linda Zagzebski, pp. 49–62. New York: Oxford University Press.

Sosa, Ernest. 2003. The Place of Truth in Epistemology. In *Intellectual Virtue: Perspectives from Ethics and Epistemology*, edited by Michael DePaul and Linda Zagzebski, pp. 155–179. New York: Oxford University Press.

Sparrow, Robert. 1999. The Ethics of Terraforming. *Environmental Ethics* 21: 227–245.

Sparrow, Robert. 2015. Terraforming, Vandalism and Virtue Ethics. In *Commercial Space Exploration: Ethics, Policy and Governance*, edited by Jai Galliott, pp. 161–178. Dorchester, UK: Ashgate.

Spudis, Paul. 2001. The Case for Renewed Human Exploration of the Moon. *Earth, Moon, and Planets* 87: 159–171.

Steinberg, Alan. 2013. Influencing Public Opinion of Space Policy: Programmatic Effects Versus Education Effects. *Astropolitics* 11: 187–202.

Sterrett, Susan. 2014. The Morals of Model-Making. *Studies in History and Philosophy of Science Part A* 46: 31–45.

Stoner, Ian. 2017. Humans Should Not Colonize Mars. *Journal of the American Philosophical Association* 3: 334–353.

Sun, Hua-Ping, et al. 2018. Natural Resource Dependence, Public Education Investment, and Human Capital Accumulation. *Petroleum Science* 15: 657–665.

Swami, Viren, et al. 2009. The Truth Is Out There: The Structure of Beliefs About Extraterrestrial Life Among Austrian and British Respondents. *The Journal of Social Psychology* 149: 29–43.

Szocik, Konrad, et al. 2018. Biological and Social Challenges of Human Reproduction in a Long-term Mars Base. *Futures* 100: 56–62.

Taylor, Paul. 1981. The Ethics of Respect for Nature. *Environmental Ethics* 3: 197–218.

Tobie, G., et al. 2014. Science Goals and Mission Concept for the Future Exploration of Titan and Enceladus. *Planetary and Space Science* 104: 59–77.

Tomalty, Jesse. 2014. The Force of the Claimability Objection to the Human Right to Subsistence. *Canadian Journal of Philosophy* 44: 1–17.

Traphagan, John. 2019. Which Humanity Would Space Colonization Save? *Futures* 110: 47–49.

Turner, Frederick. 1921. The Significance of the Frontier in American History. In Frank Turner, *The Frontier in American History*, pp. 1–37. New York: Henry Holt and Company.

Turrini, Diego, et al. 2014. The Comparative Exploration of the Ice Giant Planets with Twin Spacecraft: Unveiling the History of Our Solar System. *Planetary and Space Science* 104: 93–107.

Valdman, Mikhail. 2009. A Theory of Wrongful Exploitation. *Philosopher's Imprint* 9: 1–14.

Wang, Eric, et al. 2004. The Genetic Architecture of Selection at the Human Dopamine Receptor D4 (DRD4) Gene Locus. *American Journal of Human Genetics* 74: 931–944.

Webster, Christopher, et al. 2018. Background Levels of Methane in Mars' Atmosphere Show Strong Seasonal Variations. *Science* 360: 1093–1096.

Weinstein, J. 2009. Democracy, Individual Rights and the Regulation of Science. *Science and Engineering Ethics* 15: 407–429.

White, Frank. 2014. *The Overview Effect: Space Exploration and Human Evolution*. 2nd ed. Reston, VA: American Institute of Aeronautics and Astronautics.

White, Wayne. 2002. The Legal Regime for Private Activities in Outer Space. In *Space: The Free-Market Frontier*, edited by Edward Hughes, pp. 83–111. Washington, DC: Cato Institute.

Whiting, Daniel. 2012. Epistemic Value and Achievement. *Ratio* 25: 216–230.

Whittlesey, Robert, Sebastian Liska, and John Dabiri. 2010. Fish Schooling as a Basis for Vertical Axis Wind Turbine Farm Design. *Bioinspiration & Biomimetics* 5: 1–6.

Wigner, Eugene. 1960. The Unreasonable Effectiveness of Mathematics in the Natural Sciences. *Communications in Pure and Applied Mathematics* 13: 1–14.

Wilholt, Torsten. 2006. Scientific Autonomy and Planned Research: The Case of Space Science. *Poiesis and Praxis* 4: 253–265.

Wilholt, Torsten. 2010. Scientific Freedom: Its Grounds and Their Limitations. *Studies in History and Philosophy of Science* 41: 174–181.

Wilks, Anna. 2016. Kantian Foundations for a Cosmocentric Ethic. In *The Ethics of Space Exploration*, edited by James S. J. Schwartz and Tony Milligan, pp. 181–194. New York: Springer.

Williams, Lynda. 2010. Irrational Dreams of Space Colonization. *Peace Review: A Journal of Social Justice* 22: 4–8.

Williamson, Mark. 2006. *Space: The Fragile Frontier*. Reston, VA: American Institute of Aeronautics and Astronautics.

Wingo, Dennis. 2016. Site Selection for Lunar Industrialization, Economic Development, and Settlement. *New Space* 4: 19–39.

Yanal, Robert. 1999. *Paradoxes of Emotion and Fiction*. University Park: Penn State University Press.

Zagzebski, Linda. 1996. *Virtues of the Mind*. New York: Cambridge University Press.

Zagzebski, Linda. 2003. The Search for the Source of the Epistemic Good. *Metaphilosophy* 34: 12–28.

Zagzebski, Linda. 2004. Epistemic Value and the Primacy of What We Care About. *Philosophical Papers* 33: 353–377.

Zientek, M. L., et al. 2017. Platinum-Group Elements. In *Critical Mineral Resources of the United States—Economic and Environmental Geology and Prospects for Future Supply*, edited by K. J. Shulz et al., pp. N1–N91. U.S. Geological Survey Professional Paper 1802-N.

Zimmerman, Michael. 2001. *The Nature of Intrinsic Value*. Lanham, MD: Rowman and Littlefield.

Zubrin, Robert. 2000. *Entering Space: Creating a Spacefaring Civilization*. New York: Tarcher/Putnam.

Zubrin, Robert. 2009. The Moon-Mars Initiative: Making the Vision Real. *Futures* 41: 541–546.

Zubrin, Robert. 2011. *The Case for Mars: The Plan to Settle the Red Planet and Why We Must*. Rev. ed. New York: Free Press.

INDEX

dopamine D4 receptor (DRD4). *See* exploratory behavior

Douglas, Heather, 95–96, 101–2, 112

duties to future generations. *See* duty to extend human life; space settlement

duty to extend human life, 5, 44–45, 183–86
  and the Non-Identity problem, 187–89, 200–2
  objections to, 186–94
  rationales for, 186–94
  relationship to space settlement (*see* space settlement: relationship to duty to extend human life)
  and the value of humanity, 189–91

duty to improve human well-being, 5, 42–44, 157, 158–62, 172–75, 183
  relationship to space development (*see* space development: relationship to the duty to improve human well-being)

Earth observation from space, 108–9, 112. *See also* space science

educational inspiration. *See* scientific literacy; STEM education

Ehrenfreund, Pascale, 179–80

Elgin, Catherine, 79–83

Elvis, Martin, 165, 167, 168–69, 178–79, 180

environmental protection. *See* intrinsic value; planetary protection; space development; space science; space settlement

epistemic value. *See* Elgin, Catherine; intrinsic value: of true belief; knowledge: intrinsic value of; Kvanvig, Jonathan; Pritchard, Duncan; understanding: intrinsic value of; Zagzebski, Linda

ethical duties
  nature of/conditions for, 12–14, 183, 204
  See also duty to extend human life; duty to improve human well-being; planetary protection; scientific knowledge and understanding; space development; space science; space settlement

ethics
  obligations (*see* ethical duties)
  of space mining (*see* space development)
  of space settlement (*see* space settlement)

evolution and public opinion, 25–26, 238*t*

exploitation
  of humans (*see* space settlement: and human exploitation)
  of space resources (*see* asteroids; lunar resources; space development)

exploration. *See* curiosity; exploratory behavior; scientific exploration

exploratory behavior
  relationship to curiosity, 34–35
  relationship to space exploration, 6, 12, 32–34, 35, 37–38
  scientific characterization and explanation of, 6, 32–38

extraterrestrial life
  public opinion about, 6, 28–32
  value of (*see* intrinsic value: of extraterrestrial life; intrinsic value: of microbial life)
  value of searching for, 6, 12, 23–32
  See also astrobiology; space science

freedom of research. *See* scientific freedom

frontier metaphor, 6, 12, 38–41

funding
  for NASA (*see* National Aeronautics and Space Administration [NASA])
  for spaceflight programs, 1–2, 11–12, 106–7
  effect on degree conferral rates, 15–19, 211–39 (*see also* STEM education)

Grimm, Stephen, 63, 73n24, 78n32, 86n36

Guston, David. *See* Brown, Mark

Hickman, John, 158–59

Horwich, Paul, 62–63

human space exploration. *See* crewed versus robotic space exploration; duty to extend human life; exploratory behavior; space development; space settlement

Milligan, Tony, 9, 124, 138–39, 144–48, 172–73, 178–79, 193n13, 202n19

Moon Agreement, 159–61, 175–77, 181. *See also* lunar resources; Outer Space Treaty; space development

Moore, G.E. *See* intrinsic value: concepts of

moral obligation. *See* duty to extend human life; duty to improve human well-being; ethical duties; planetary protection; scientific knowledge and understanding; space science

Mousis, Olivier, 108–9, 113–14

Munévar, Gonzalo, 89–91, 91n4, 92–96, 103, 107–8, 192n12

National Aeronautics and Space Administration (NASA), 1, 11–12, 15–16, 17, 18, 19, 91
  funding for, 1–2, 11–12, 15–19, 26–28, 27t, 211–36, 238f, 239f, 240t
  impact on scientific literacy (*see* scientific literacy)
  impact on STEM education, 15–19, 21–23, 211–36 (*see also* STEM education)
  and public opinion, 21–23, 22f, 26–28, 27t, 238f, 239f, 240t

National Space Society (NSS), 38, 40n31, 41, 42

near-Earth asteroids. *See* asteroids

near-Earth objects. *See* asteroids

Non-Identity problem. *See* duty to extend human life: and the non-identity problem

novelty-seeking. *See* exploratory behavior

O'Neill, John, 132–33

orbital allocation regulations, 175–76

Outer Space Treaty, 2, 8–9, 122–23, 156, 159–60, 161, 175–76. *See also* planetary protection; space development

Peaks of Eternal Light. *See* lunar resources

permanently shadowed craters. *See* lunar resources

Pettinico, George, 28–31

planetary protection
  Committee on Space Research's (COSPAR) Planetary Protection Policies, 122–23, 124–28
  for interstellar missions, 124, 152–54
  for lifeless space environments (*see* intrinsic value: of lifeless space environments; planetary protection: rationale for expanding policies)
  rationale for COSPAR's policies, 7, 126–28 (*see also* intrinsic value: of extraterrestrial life)
  rationale for expanding policies, 7–8, 120–21, 123–24, 137–54, 175–82, 194–97
  relationship to scientific exploration, 8, 109n17, 115, 124–28, 148–52
  See also space science: conflicts with space development; space science: conflicts with space settlement

planetary science, 108–10, 113–15, 117–19, 140, 143–44, 148–52, 163, 165–70, 177, 179–80, 195–97, 207. *See also* space science

platinum-group metals. *See* asteroids

Pop, Virgiliu, 160–61, 173–74

Pritchard, Duncan, 66–83. *See also* understanding

private space exploration. *See* space development

public opinion. *See* evolution and public opinion; extraterrestrial life; National Aeronautics and Space Administration (NASA); space science

Race, Margaret, 138, 179–80

Ratke, Lorenz, 110–11

Regis, Ed, 203–5. *See also* interstellar travel; space settlement

religion. *See* evolution and public opinion; extraterrestrial life; National Aeronautics and Space Administration (NASA)

remote sensing. *See* Earth observation from space

Rendall, Matthew, 201–2

right to research. *See* scientific freedom

Rivkin, Andrew, 167, 170

robotic space exploration. *See* crewed versus robotic space exploration; planetary protection

Rolston, Holmes, 124, 138–44

Roussos, Panos, 35–36, 37n28

Rummel, John, 138, 194–95

Sadeh, Eligar, 15–16

Sagan, Carl, 19–20, 32–33, 143

Sanchez, Joan-Pau, 167–68, 169f, 171

scientific exploration. *See* planetary protection: relationship to scientific exploration; scientific knowledge and understanding: relationship to scientific exploration

scientific freedom
    of ends, 97–120, 207–8
    of means, 97–101, 106–20, 207–8
    See also Brown, Matthew; Kitcher, Philip; scientific knowledge and understanding; space science

scientific knowledge. *See* knowledge; scientific knowledge and understanding; understanding

scientific knowledge and understanding
    instrumental value of, 3, 4–5, 7, 9–10, 12, 45–46, 91–120, 148–52, 207–8
    intrinsic value of, 3, 4–5, 6–7, 9–10, 45–46, 50–53, 56, 64, 82–88
    obligation to acquire, 3–6, 7, 10, 45–46, 47–48, 50–53, 83–88, 89–91, 97–120, 124, 146–52, 194–97, 206–9
    relationship to democratic governance, 7, 89–91, 97–120, 206–8
    relationship to scientific exploration, 7, 89–90, 91–97, 106–20, 148–52
    relationship to scientific progress, 3, 7, 89–90, 91–97, 106–20, 148–52
    relationship to societal progress, 3, 7, 89–97, 208
    varieties of (pure or basic research vs. applied research or technology development), 90–91, 95–96, 102–20
    See also knowledge; planetary protection; scientific freedom; space science; understanding

scientific literacy, 6, 12, 15, 19–23. *See also* STEM education

scientific progress. *See* scientific knowledge and understanding: relationship to scientific progress

scientific understanding. *See* scientific knowledge and understanding; understanding

search for extraterrestrial life. *See* astrobiology; extraterrestrial life

serendipity of science. *See* Brown, Matthew; Kitcher, Philip; Munévar, Gonzalo; scientific knowledge and understanding; space science

Shelhamer, Mark, 119–20

significant questions. *See* Brown, Matthew; Kitcher, Philip

Smith, Kelly, 133–37, 199–200, 202–3

Solomon, Lewis. 160n4, 161, 173

Sosa, Ernest, 57–58

space colonization. *See* space settlement

space development
    conflicts with space science (*see* space science: conflicts with space development)
    objections to, 6, 8–9, 42–44, 171–82
    rationales for, 1–2, 8–9, 12, 38–41, 42–44, 158–62
    relationship to the duty to improve human well-being, 5, 8–9, 12, 42–44, 157, 158–62, 171–82
    See also asteroids; lunar resources

space frontier. *See* frontier metaphor

space law. *See* asteroids; Commercial Space Launch Competitiveness Act of 2015; lunar resources; Moon Agreement; Outer Space Treaty; planetary protection; space development

space mining. *See* asteroids; lunar resources; space development

space resources. *See* asteroids; lunar resources; space development

space resource exploitation. *See* asteroids; lunar resources; space development

space science
    conflicts with space development, 3–6, 8–9, 10, 42–43, 120–21, 156–58, 175–82, 209

space science (*cont.*)

    conflicts with space settlement, 9, 45, 120–21, 184, 194–97, 209

    obligation to conduct, 3–6, 7, 10, 14–15, 47–48, 88, 89–92, 97, 106–20, 148–52, 157–58, 178–82, 194–97, 206–9

    and public opinion, 22–23, 26–28, 27*t*, 237–40

    relationship to democratic governance, 7, 10, 89–91, 97, 106–20, 206–8

    See also planetary protection; scientific freedom; scientific knowledge and understanding

space settlement

    basic argument for, 9, 183–84, 185–94

    conditions of life within, 9, 39, 197–205, 206–8

    conflicts with space science (*see* space science: conflicts with space settlement)

    and cultural diversity, 40–41

    and the disposable planet mentality, 191–92

    and governance issues, 39, 99–100n12, 199–200, 206–8

    and human exploitation, 9, 184, 200–5

    and informed consent, 197–199

    objections to, 9, 39–41, 186–205

    rationales for, 6, 32–34, 38–41, 44–45, 184–86

    relationship to the duty to extend human life, 5, 9, 12, 44–45, 153–54, 183–94, 199–205

    and reproductive autonomy, 202–5, 207

    urgent argument for, 9, 183–84, 185–86, 193, 194–97, 205

species survival. *See* duty to extend human life; space settlement

spending on spaceflight. *See* funding: for spaceflight programs; National Aeronautics and Space Administration (NASA): funding for

spinoffs. *See* scientific knowledge and understanding: instrumental value of

Spudis, Paul, 116, 118–19

STEM education, 2, 6, 12, 13–14, 15–19, 211–36. *See also* scientific literacy

sustainable uses of the space environment. *See* asteroids; lunar resources; planetary protection; space development; space science; space settlement

swamping problem, 65–66, 67–68, 71. *See also* knowledge; understanding

Taylor, Paul, 130–31, 132, 134–35

Tobie, G., 114–15

true belief. *See* intrinsic value: of true belief

understanding

    conception of, 68–69, 79–81

    instrumental value of, 7

    intrinsic value of, 6–7, 48, 54–55, 64, 66–67, 69–79, 81–83, 85–88

    See also knowledge; Elgin, Catherine; Pritchard, Duncan; scientific knowledge and understanding

virtue epistemology. *See* Baril, Anne; Zagzebski, Linda

virtue ethics, 83–88. *See also* Baril, Anne; Linda Zagzebski

water. *See* asteroids; lunar resources

well-ordered science. *See* Brown, Matthew; Kitcher, Philip

Wilholt, Torsten, 97, 101n13, 103, 106–7, 110

Williams, Lynda, 191

worldship travel. *See* interstellar travel; space settlement

Zagzebski, Linda, 64, 65n15, 84–85, 88

Zubrin, Robert, 32, 40–41